Mineral Resource Base of the Southern Caucasus and Systems for its Management in the XXI Century

NATO Science Series

A Series presenting the results of scientific meetings supported under the NATO Science Programme.

The Series is published by IOS Press, Amsterdam, and Kluwer Academic Publishers in conjunction with the NATO Scientific Affairs Division

Sub-Series

I. Life and Behavioural Sciences	IOS Press
II. Mathematics, Physics and Chemistry	Kluwer Academic Publishers
III. Computer and Systems Science	IOS Press
IV. Earth and Environmental Sciences	Kluwer Academic Publishers
V. Science and Technology Policy	IOS Press

The NATO Science Series continues the series of books published formerly as the NATO ASI Series.

The NATO Science Programme offers support for collaboration in civil science between scientists of countries of the Euro-Atlantic Partnership Council. The types of scientific meeting generally supported are "Advanced Study Institutes" and "Advanced Research Workshops", although other types of meeting are supported from time to time. The NATO Science Series collects together the results of these meetings. The meetings are co-organized bij scientists from NATO countries and scientists from NATO's Partner countries – countries of the CIS and Central and Eastern Europe.

Advanced Study Institutes are high-level tutorial courses offering in-depth study of latest advances in a field.
Advanced Research Workshops are expert meetings aimed at critical assessment of a field, and identification of directions for future action.

As a consequence of the restructuring of the NATO Science Programme in 1999, the NATO Science Series has been re-organised and there are currently five sub-series as noted above. Please consult the following web sites for information on previous volumes published in the Series, as well as details of earlier sub-series.

http://www.nato.int/science
http://www.wkap.nl
http://www.iospress.nl
http://www.wtv-books.de/nato-pco.htm

Series IV: Earth and Environmental Sciences – Vol. 17

Mineral Resource Base of the Southern Caucasus and Systems for its Management in the XXI Century

edited by

Alexander G. Tvalchrelidze

Chairman of the Board,
Union 'Georgian Resources and Sustainable Development',
Tbilisi, Georgia

and

Georges Morizot

Director for Europe and Western Asia,
Bureau de Recherches Géologiques et Minières (BRGM),
Orléans, France

Kluwer Academic Publishers

Dordrecht / Boston / London

Published in cooperation with NATO Scientific Affairs Division

Proceedings of the NATO Advanced Research Workshop on
Mineral Resource Base of the Southern Caucasus and Systems for its Management in the XXI Century
Tbilisi, Georgia
3–6 April 2001

A C.I.P. Catalogue record for this book is available from the Library of Congress.

ISBN 1-4020-1123-7

Published by Kluwer Academic Publishers,
P.O. Box 17, 3300 AA Dordrecht, The Netherlands.

Sold and distributed in North, Central and South America
by Kluwer Academic Publishers,
101 Philip Drive, Norwell, MA 02061, U.S.A.

In all other countries, sold and distributed
by Kluwer Academic Publishers,
P.O. Box 322, 3300 AH Dordrecht, The Netherlands.

Printed on acid-free paper

Printed in the Netherlands.

TABLE OF CONTENTS

Preface

The NATO Advanced Research Workshop "Mineral Resource Base of the Southern Caucasus and Systems for its Management in the XXI Century" was held in Tbilisi on April 3-6, 2001. The workshop was financed by the NATO Science Programme and financially supported by the Open Society – Georgia Foundation.

By the end of the XX century the new states of the Southern Caucasus (Azerbaijan, Armenia and Georgia) found themselves in the lowest phase of a severe structural crisis. Belonging to the same geopolitical space and having the common economic history during the XIX and XX centuries, these countries will be able to overcome this crisis and to integrate into the world community only step-by-step, exploiting their resource bases and developing competitive branches of industries. Moreover, in our opinion, such conception is the only alternative to the spontaneous economical chaos. Solution of this problem seems to be impossible without close international collaboration. From this point of view, the idea of regional alliances where regional actors would be linked by a general, non-contradictory macroeconomic model seems to be the only possibility for sustainable development. Creation of The Southern Caucasus Alliance (SCA) where Georgia, Azerbaijan, Armenia and Turkey would find their place could be possible only in case if the economic reality and economic interests would weigh down nationalistic doctrines of political isolationism. Such approach may become ultimately realistic if it would take into consideration the most important national interests of all the actors and forward a macroeconomic strategy acceptable for all countries. The further macroeconomic development as the only guarantee of regional security would greatly depend on the resource base of the whole regional geopolitical space as well as on synergetic macroeconomic model determining non-contradictory integration of the actors into the world community.

Elaboration and implementation of a long-term state programmes for sustainable macroeconomic development of the Southern Caucasus countries in the XXI century depends of three major factors:

- Thorough analysis of resource bases using the modern adequate methods, original computer software and corresponding data banks and information systems;
- Comparative analysis of the main trends of economic development based on national resource bases.

The Workshop held in Tbilisi was the first attempt to discuss the role of mineral resources in the economy of the XXI century and to find common approaches for further international collaboration. The Workshop was attended by scientists and specialists not only from Turkey, Azerbaijan, Armenia, Russia and Georgia, but also from France, UK, Greece, Portugal, Albania, Germany, Czech Republic and Ukraine. Within the last 12 years this Workshop was the first international event, which allowed the specialists to renew contacts with colleagues from different countries, obtain new information, analyse development tendencies of mineral resources and have a vis-à-vis informal discussions.

It was decided to organize permanent workshops. It was also decided to publish Proceedings. Unfortunately, for more than a year we were unable to collect necessary articles and are publishing now only those materials we had a chance to receive before July 17, 2002. We performed only slight editing of the received materials preserving writing style of each author and maintained mutual layout for the whole volume.

Alexander G. Tvalchrelidze
Georges Morizot

SUSTAINABLE DEVELOPMENT IN THE NON-ENERGY EXTRACTIVE INDUSTRY

Southern Caucasus case
The European Union Approach
Some tools for the Promotion of Sustainable Development in the Non-Energy Extractive Industry

G. MORIZOT
BRGM- International Division
B.P. 6009- 45060 ORLEANS CEDEX 2
France
E-mail: g.morizot@brgm.fr

1. Introduction

The term 'sustainable development' corresponds to a concept that has been, during the last decade, increasingly in use in different circles (scientists, economists, politicians, bankers, etc.). No development nowadays seems to be worth to be considered if it could not be qualified as sustainable. Now to the word "sustainability" is generally associated the concept of duration and permanent progress: how is it possible to associate such concept of duration and permanent progress to the exploitation of natural resources which are not renewable?

How could the exploitation of mineral resources correspond to the concept of sustainable development, when any professional, working in the extractive industry knows that the ore body in exploitation will be exhausted more or less quickly, leading to the closure of the industrial exploitation?

We will try in the present paper:

- To define more precisely the term 'sustainable development'.
- To demonstrate in which conditions the non-energy extractive industry could fulfil the conditions of the sustainable development.
- To present what is the corresponding European approach to the sustainable development.
- To present some technical tools that can support the concept of sustainable development both for the Administration and for the Mining Industry.

A.G. Tvalchrelidze and G. Morizot (eds.),
Mineral Resource Base of the Southern Caucasus and Systems for its Management in the XXI Century, 1–18.

2. Sustainable development and the non-energy extractive industry

2.1. DEFINITION OF THE SUSTAINABLE DEVELOPMENT

The sustainable development - as defined by the UN Bruntland Commission on Environment & Development has "to meet the needs of the present generation without compromising the ability of future generation to meet their own needs"- is founded on an equilibrium between three pillars, the economic (competitiveness), the environmental, and the social: to comply with the sustainable development, "all businesses can and must achieve the three interlinked goals of economic prosperity, environmental protection and social equity" as explained by Elkington John [7].

Figure 1. The three pillars of sustainable development

Until recently the terms "mining industry" and "sustainable development" have been presented as incompatible, taking into consideration the depletion of an ore body after its exploitation. It is obvious that the exploitation of non-energy natural resources in many countries of the world and between them the former Soviet Union could not be considered as satisfying the concept of "sustainable development" taking into consideration the impact on the environment and the social situation.

2.2. MINING INDUSTRY AND SUSTAINABLE DEVELOPMENT

In fact, it is now demonstrated by Prévôt et al. [13] that the mining industry could fulfil the conditions of sustainable development taking into consideration different approaches. For example, the sustainability should be considered as "spread over several generations and the wealth produced by the use of non-renewable resources (natural capital) is transferred onto the next generations in the form of human capital development (the case of Europe's transition from XX century industrial development to the late XX century post-industrial society described by Owens in 1994 [12].

The sustainability could then be established as follows:

- On a world-wide scale, in developing new resources (new mines) or substitutes to replace resources currently being used, and
- At a specific site, if the resources are used as efficiently as possible while meeting the community's social goals and, if in a second stage, activities of substitution are found and established in order to take over from the mining after the exhaustion of the mineral reserves.

It could be said that the sustainable development implies the substitution of depleted resource by an investment in human capital development and infrastructure development which can contribute both to the environmental sustainability and the social development. Sustainable development requires a rational extraction of resources, starting with the easier reserves, enabling accumulation of experience and improvement of technologies [12].
As a consequence, in order to implement the accepted principle of sustainability, **appropriate regulations, company strategies, and industrial and governmental structures are needed, together with an adequate land-use management allowing an efficient planning of mining resources.**

2.3. MULTIPLYING AND DEVELOPMENT EFFECT OF THE MINING INDUSTRY

In France, Germany, UK the Mining Industry has been the "mother" Industry for the development of the modern Industry, thanks to the energy and raw materials produced, to the infrastructures developed around the mining industry, thanks also to the availability of skilled manpower trained in mining industry.
To emphasize the importance of the non-energy extractive industry for the economy of a country some figures, presented by Cliff, should be given on its multiplying effect [4]:

- The creation of one job in mining industry is generating four jobs in the support industries and services (metallurgy, chemical industry, transport, etc.).
- The revenues of mining industry for a given country could be eight times the direct revenue generated by taxes.

3. Southern Caucasus and mining industry

The Southern Caucasus region has a long tradition of the exploitation of mineral resources:

- There exists evidence of copper mining in Armenia since 6000 BC.
- The origin of the legend of the Golden Fleece, searched by the Argonauts in Colchis (the territory presently watered by the River Rioni, near Poti, in the modern Georgia), is directly linked to the method of extraction of alluvial gold trapped on sheep skins, following Graves [9].
- Around 3000 BC the area was supplying copper, gold and silver to the Mediterranean countries.
- In the X Century mining of massive sulphide is recorded.
- During the modern period, mostly in the XX century, mining was far from fitting the

conditions of the sustainable development: the mining of the metallic ores and the construction materials had been developed at the time to service the societal and industrial needs of the Soviet Union. The two pillars corresponding to the competitiveness and environmental protection were not taken into consideration. The region was known then for the metallic production as follows: iron and aluminium in Azerbaijan, copper, molybdenum, zinc and gold in Armenia, manganese, copper and aluminium in Georgia.

Nowadays, in the present context where economical, social and ecological rules have been dramatically modified, the extractive industry in spite of suffering severely remains in most of the countries a pillar of the economical development. The importance of the mining sector has been emphasized by another presentation in the frame of the Workshop "Mineral Resource Base of the Southern Caucasus and Systems for its Management in the XXI Century".

Evidently, the situation from one country to another, and inside one country from one industrial site to another can be, and really is, very different.

Between the negative factors, we should mention:

- The destruction of the economical and industrial network and labour division system (geographical distribution) existing in the former Soviet Union and the decrease of the national markets (e.g. military orders).
- Emergence of environmental constraints.
- Frequent lack of adapted regulatory and taxation systems or, when developed, frequent lack of enforcement capabilities (problem of the adaptation of the administration to its new role).
- Obsolete industrial technologies and outdated equipment.
- Sharp decrease of traditional exploration efforts
- At regional scale: conflicts, creation of new barriers, energy blockade, and insufficiency of regional co-operation.
- Inadequate access to the information.
- Weight of the past (industrial, social, and environmental).
- Lack of the national capital and difficulties in attracting foreign capital.

On the other hand, the development of mining activity inside Central Asian countries benefits from positive factors:

- Richness of the sub-soil in mineral resources.
- Openness of the world economy for goods and capital.
- Importance of past geological work and accumulated exploration data.
- Qualification and low cost of the manpower (engineers, geologists, etc.).
- Expected growth of internal markets, namely for industrial minerals and construction materials in chemical, petrochemical, building industry.
- Definition of new regulatory framework.

As demonstrated by the boom of mining exploration in Scandinavian countries and specifically in Sweden, after the adaptation of the mining legislation, the extractive industry is not an industry of the past reserved to developing countries that are not much concerned by ecological problems: availability of metals is a need for the whole mankind and the application of the concept of sustainable development to the mining industry allows to conciliate mining activity, social development, and environment protection.

4. The European Union and sustainable development in mining industry

Nowadays, the concept of sustainable development inside the 15 members of the European Union is the object of proposition and intense exchanges between different stakeholders:

- The governments and the European Commission (different Directorates in charge of the enterprises, the protection of the environment, the competition, etc.).
- Different mining enterprises and their national and European federations, Euromines.
- The NGO, as WWF.

At worldwide level it is also the object of an initiative taken by the world's major mining companies, known as Global Mining Initiative (GMI). See the site www.globalmining.com.
The sustainable development of the Industry, including the environment protection policies, has been a concern of the European Community, now the European Union since its creation (article 6 of the Treaty establishing the former European Community). The non-energy mining industry was the first to become the specific object of a Communication from the Commission [5]. Its "objective is to set the broad policy lines for promoting sustainable development in the EU non-energy extractive industry by reconciling the need for more secure and less polluting extractive activities while maintaining the competitiveness of the industry".
Taking into consideration the fact that with the exception of the renewable resources derived from agriculture, all raw materials used for everyday life are extracted from the earth or the sea, and that the EU is the world's largest consumer of minerals, the EU considers that the sustainability of corresponding industry is necessary to preserve the resources of the future generations and also to minimize the impact of mining on the environment.

4.1. PRIORITY ISSUES FOR THE EU

They are the following:

- The high level of environment protection wanted by the EU requires an improvement of the legislative framework and also the development of new instruments for, as examples, the management of mining wastes, the implementation of key principles such as polluter- pays, prevention and precaution, the adequate restoration of the environment, the development of environmental agreements and of voluntary initiatives fostering improvements in environmental performance.
- The development of the industry requires an access to the geological deposits which determines the location of the extractive operations: the need for land access for

exploration and mining must then be balanced with a high level environment protection and the existence of protected areas.

- The maintaining of a business climate conducive to investment in the industry requires the existence of adequate procedures for the right to explore and to exploit. At the same time the competitiveness of the European industry versus the industry of the rest of the world requires a benchmarking of the different legislation existing in the main mining countries.
- A high level of the efforts in the field of Research and Technological Development is necessary to better evaluate and manage the impacts on the environment, to develop new technologies (e.g. information technologies), to improve the competitiveness and the environmental and social performance of the industry: a better integration and co-operation of the research efforts inside the EU is recommended by the Commission.
- The enlargement of the EU through the accession of new candidate countries, in which the mining industry is often facing heavy competitiveness, as well as environmental and social concerns, is also an important challenge for the EU.

4.2. EUROPEAN UNION ACTIONS

Between the important EU actions in the sector we should mention:

- The communication from the Commission, dated May 3rd 2000 and titled "Promoting sustainable development in the EU non-energy extractive industry" [5].
- The decision of the Council of the EU, dated June 29th 2000, recommending "building as appropriate on existing structures, in order to encourage the emergence of conditions in which all the stakeholders can co-operate more effectively towards the goal of sustainable development of the industry, in an integrated and balanced way" [6].
- A study on voluntary initiatives in industry to foster environmental performances [13].
- A general inventory of the mining wastes in the EU and in the main mining countries from CEEC.
- The definition of indicators of sustainable Development.

4.3. EUROMINES ACTIONS

Euromines, as the Association of the European Mining Industry has developed following actions:

- Definition of guidelines on sustainable development [8], including following different recommendations:
  - Access to and use of minerals resources is integral to Sustainable Development for present and future generations,
  - Members of the European mining industry shall conduct their activities according to the principles of Sustainable Development,

- Minimization of effects on the environment shall be integral to the conduct of mining activity,
- Environmental protection should be based on sound science and be technically and economically feasible,
- The industry shall offer good, safe and positive working conditions taking all necessary protections,
- The industry shall identify, minimize and prevent potential risks linked to its activity.

- Permanent participation in the EU activities related to mining, as a major stakeholder.

4.4. SUSTAINABLE DEVELOPMENT INDICATORS

We could mention that presently (summer 2001) indicators related to sustainable development are discussed in working groups including the Industry, mainly represented by Euromines, NGOs (WWF), and representative of the EU.
As examples, the following Key fields and indicators are under discussion.

Key Fields	Indicators
Land management	Rate of restored land
Natural resource management	Rate of ore recovery
Energy	Green-house emission
Water	Fresh water consumption
Health & Safety	Safety performance
Employment conditions	Training performance
Air pollution	Pollution/legal limits

5. Some examples of tools able to support the sustainable development of mining industry

We could mention the following tools, mainly based on the new information and communication technologies and on bio-processes:

- Geographic Information Systems and comprehensive decision-aiding systems.
- Modelling and simulation.

- Optimizing Control Systems.
- Bio-technologies.

5.1. GEOGRAPHIC INFORMATION SYSTEMS AND COMPREHENSIVE DECISION-AIDING SYSTEMS

The geographic information systems make it possible to store, distribute, organize and treat information, in digitalized form, relative to a geographical point defined by its co-ordinates. The information can cover different fields, such as geology, meteorology, ecology, land property, sociology, agriculture, etc. Offering the possibility to store data and to exploit the information in an easier way, they are a very valuable tool, taking into consideration the numerous parameters playing a role in the sustainable development. Their importance is recognized by the European Union. "Members states are also invited to share experience and information, for example, on balanced approaches to assigning areas of land for future extractive operations and how comprehensive decision-aiding systems, integrating data on land use, bio-diversity, cultural heritage, geology, and water resources, can be effectively developed and applied".

In the case of Southern Caucasus Region, where many valuable data accumulated through heavy Geological expeditions, not any more conceivable in the present socio-economical conditions, remain in a fragile and perishable form (handwritten or paper printed materials), the use of information systems offer the possibility to save such accumulated information in an easy-to-handle form. Based on geo-referenced, complete and reliable data banks, they make it possible to give access to a synthetic presentation of the mining potential of a country at different scales, for the benefit of the investors, the administration, the geologists and the public.

The availability of all the existing information transcript in a language corresponding to the concepts used in the International Mining Communities allows **the potential investors** to be sure to have access (for example, through the Internet or through CD-ROM, either free of charge or at a defined cost) to all pertinent data, to treat such information by modern ore bodies evaluation and projects evaluation tools (geostatistics, mining project software, etc.). In addition, they would highly appreciate the possibility to exchange electronic information with the administration.

The geologists are conscious that the discoveries of new ore bodies are more and more based on the superposition of data and the generation of simulations, taking into consideration new concepts and models: the creation of geological information systems valorizes existing data which are often very expensive to acquire and in all cases leads to a modernization of the concepts and models which have been used in the interpretation of existing data; it makes necessary a modernization of the definition of geological terminology used in the system.

The geologist will then be able:

- To use modern treatment methodologies, such as Multidata processing, Hierarchical Ascendant Classification, Main Component Analysis, etc.

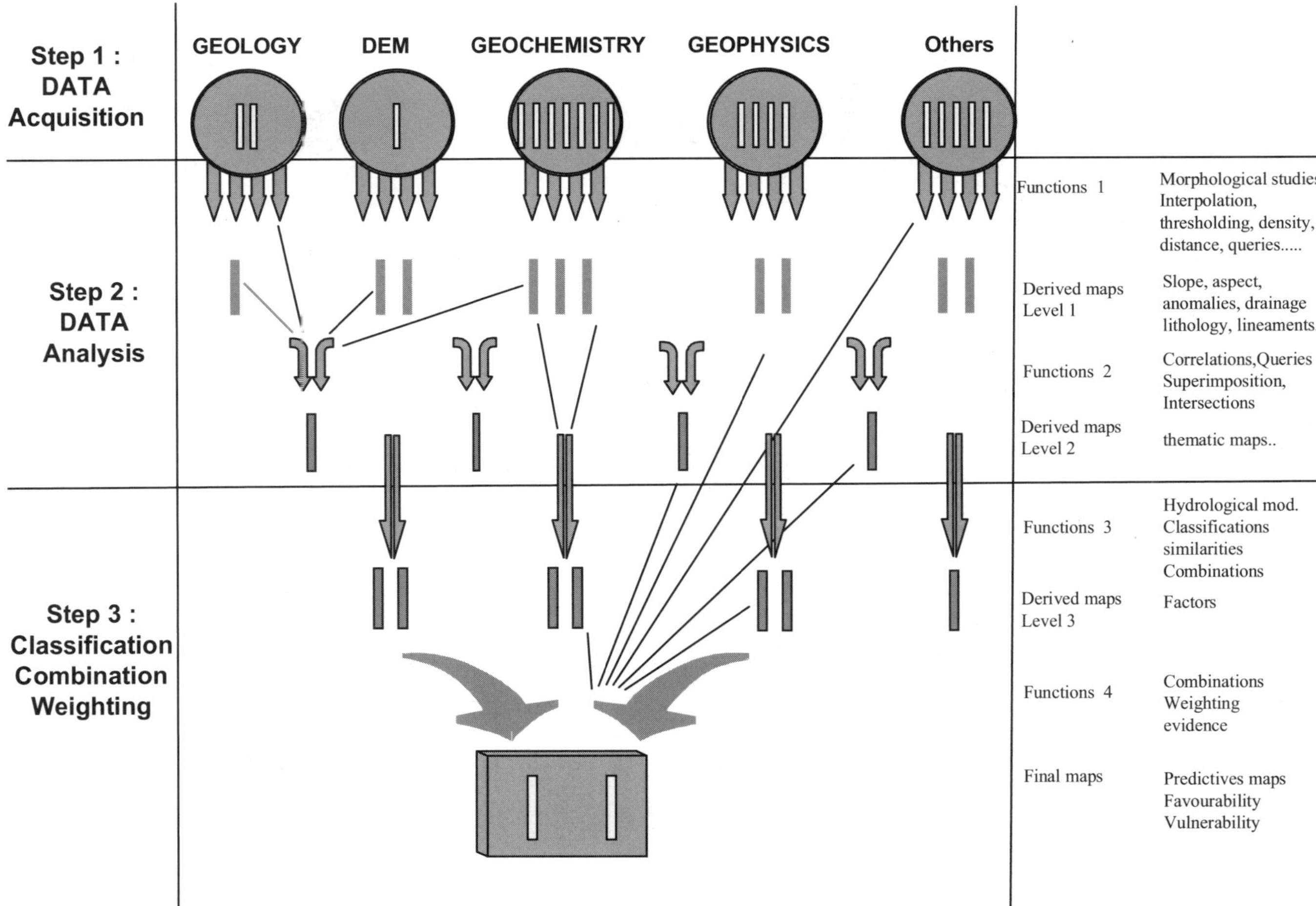

Figure 2. Data processing with SynArc [1].

- To make possible the creation of different thematic maps and decision-helping documents, useful for the Administration and industrial operators.

The availability of a data-processing tool, such as SynArc [1], produced by BRGM will make it possible:

- To use data for different sources (as satellite images or geological, geochemical, geophysical and topographic surveys).
- To combine and process data (raster and vector), creating a synergy between different data types through a large range of classifications, interpolations, morphological, drainage and statistical analysis.
- To generate modelling and simulation, for example for the visualization of areas vulnerable to pollution or landslides, or those that might contain mineral resources or would be suitable for the construction of infrastructure elements.
- To provide maps, analytical and synthetic images and tables that will help to make decisions.

The addition of environment protection related data would make it possible to model the behaviour of the pollution, to evaluate corresponding risks for the human health, and to evaluate different alternatives for sustainable development.

As example of possible environmental information, there could be mentioned:

The description of the environment:

- The cadastre of the different protected areas (as natural parks).
- The repartition and characterization of the human activities (social and economical aspects).
- The soil occupation.
- The inventory and the characterization of the pollution sources of the different activities.
- The Climatology.
- Ground and surface waters.
- Vulnerability and natural risks (erosion, earthquakes, flooding).
- Geothermal energy.
- Monitoring networks.

An information base:

- Data on potential pollutants and associated risks.
- Data on the different sectors of activity: consumables, products and sub-products.
- Regulatory framework (laws, norms and standard).

For the administration, in addition to the availability of synthetic documents able to present the mining potential of the country to possible investors, the information systems will be the a basis for the sustainable development of the territory, making it possible through different simulations and data combinations (comprehensive decision-aiding systems) to identify critical problems and to be realistic in drawing up environmental regulation. They offer a tool for the strengthening of environmental agencies and also for an approximation to European environmental legislation.

As an example of GIS use in surface and ground water pollution sensitive analysis, we could mention a study on the regional impact of Mining activities in South Africa carried out by the South African Council for Geosciences and BRGM, as described by Chevrel and Coetzee [3]. Its objective was to assess the environmental damage from the mining activity at a regional scale through the inventory of polluted areas, the definition of areas susceptible to pollution and the identification of pollution pathways. The study takes into consideration the geology, the land-use obtained through remote sensing images (Spot and Landsat satellites), the slope, the drainage (obtained trough the combination of different factors, such as the nature of the rocks, the land-use, the fracturation, etc.), and the topography. It makes possible to identify the risks and to define an assessment approach and corresponding prevention measures through the realization of different maps, defining:

- The pollution risk.
- The groundwater vulnerability.
- The potential drainage contamination.

The study demonstrates the effectiveness of the methodology for the impact assessment of planned and existing mining projects, the identification of areas likely to be affected by pollution (radio-active elements), the prioritization of more detailed investigations and clean-up activities (See on figure 3.).

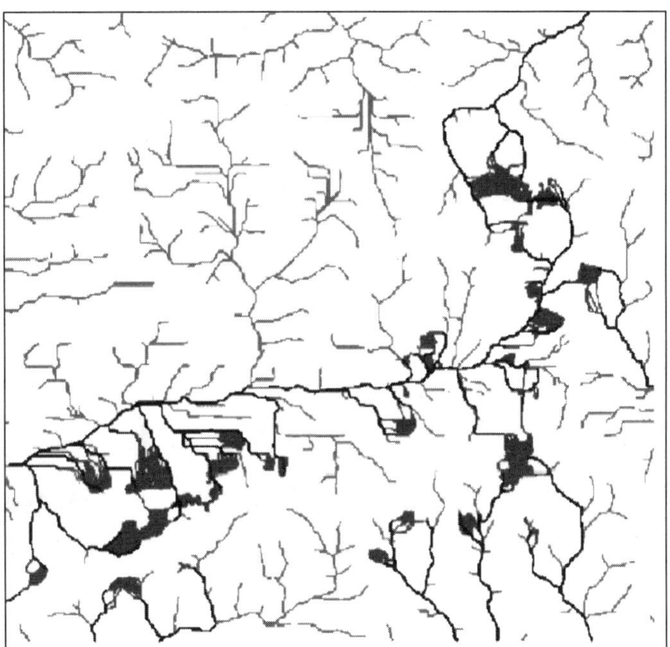

Figure 3. Potential drainage contamination map. Drainage in light grey, contaminated surfaces run-off in dark grey, contaminated drainage in black [3].

5.2. MODELLING AND SIMULATION IN MINERAL PROCESSING

We are presenting two different approaches, which can optimize the mineral processing operations, the first one, presented in this (5.2.) paragraph, assumes that the plant is working under steady state conditions and does not take into consideration e.g. the ineluctable variation of ore quality.

In the next paragraph (5.3.) there will be presented another approach based on a permanent reaction to the variation of the system.

The modern tools of modelling and simulation, as USIM PAC described by Guillaneau et al. [10], helping to better design new operations and improve existing ones, are very valuable tools for the sustainable development of industrial operations. They are able, for defined ore characteristics, to:

- Help in the analysis of the performance of the different components of a plant.
- Optimize the plant and its global performance.
- Better control the environmental impacts.
- Easily scale up new operations.

USIM PAC includes the basic functions required to conduct a design or an optimization project:

- Capabilities for data input and results display.
- A material balance algorithm to validate and reconcile data.
- Direct simulation to predict all plant stream description from the feed characteristics and the equipment parameters.
- Reverse simulation to adjust model parameters or back-calculate optimal sizes and settings of a required configuration (see figure 4).
- Capital cost estimation.

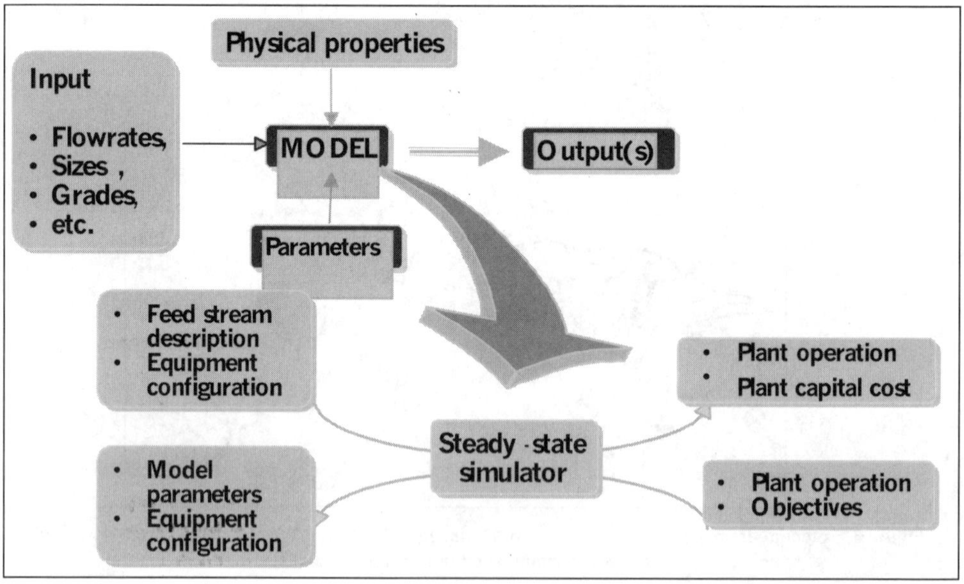

Figure 4. Basis of simulation [10]

Figure 5 shows the methodology to make the preliminary design of a new plant. The first steps use feed characteristics and main objectives to define the future plant requirements in terms of flowsheet and stream descriptions. The next step back-calculates with reverse simulation the sizes of the main pieces of equipment. The last step simulates the future plant operations and calculates the cost investment. These results allow the process engineer to compare several flowsheets according both to their technical and economical performances.

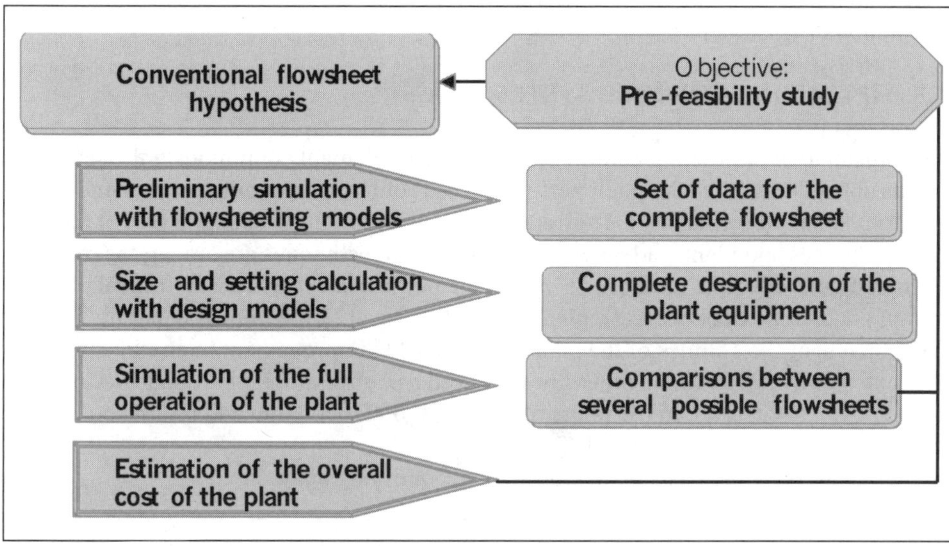

Figure 5. Design methodology [10]

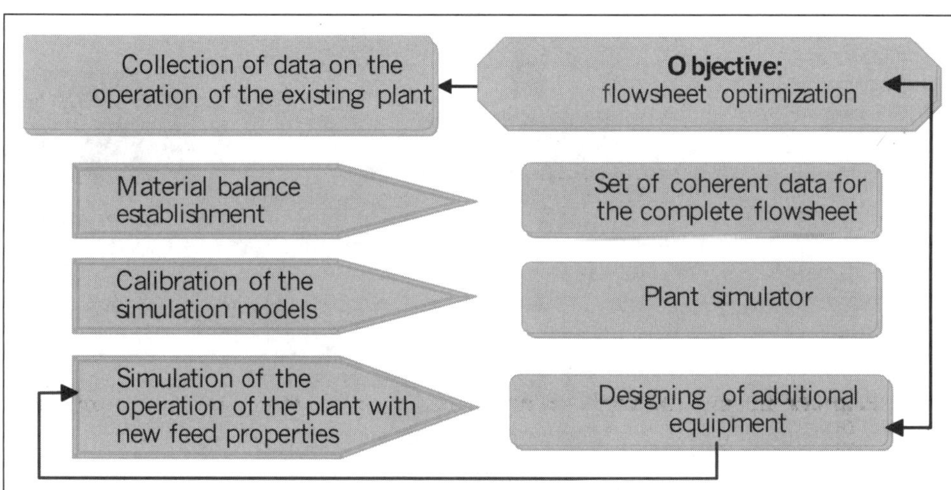

Figure 6. Adaptation methodology [10]

Figure 6 shows simplified methodologies to adapt an existing plant to new objectives or constraints. The methodology is divided into two main steps:

- The first consists in creating a simulator of the existing processing plant.
- The second — in testing different modifications with this simulator and evaluating all the simulation results in technical (characteristics of the products, power drawn by the main equipment), and economic (estimation of the capital cost investment) terms.

Methodologies involving model calibration steps are used systematically for the adaptation of existing plants.

5.3. OPTIMIZING CONTROL SYSTEMS (OCS)

The objective of optimizing control systems is to implement continuously the sets of points able to maximize the plant performances (both peak economy performances and environmental performances). They are nowadays becoming an integral part of modern mineral processing plants: advanced process control techniques are among the most cost- and time-effective methods of improving performances of a metallurgical plant. Among numerous examples, at a plant in South Africa [2], the implementation of an OCS on one of the 15 lines of the plant led to an 11% improvement in energy efficiency. The extension in 2000 to the other lines allowed a payback time of around two to four weeks. Typical costs are in the range of US \$ 100 000 – 500 000 and returns on investment typically range from 100 to 5000 %.
Figure 7. presents the basis of an Optimizing Control System.

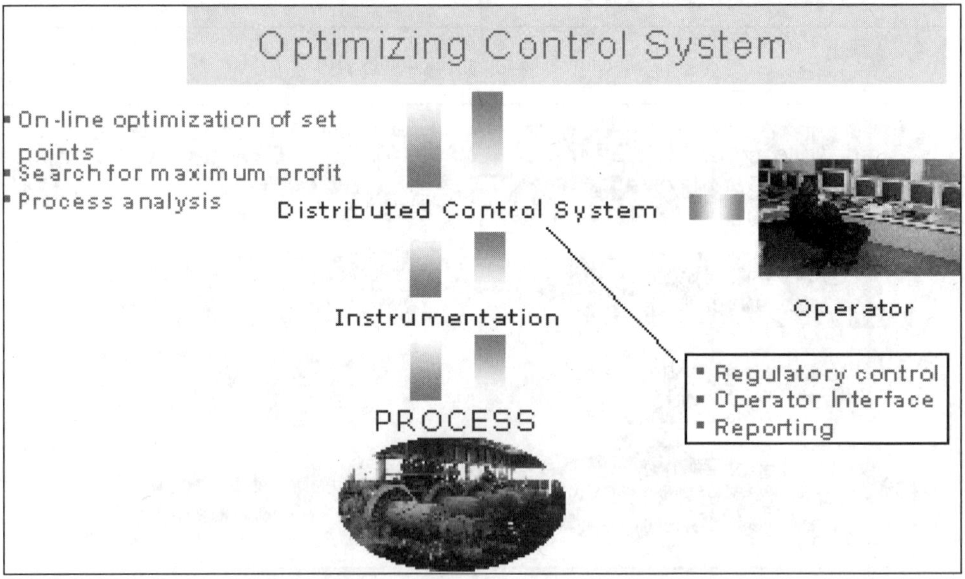

Figure 7. Basis of an Optimizing Control System (OCS) [2]

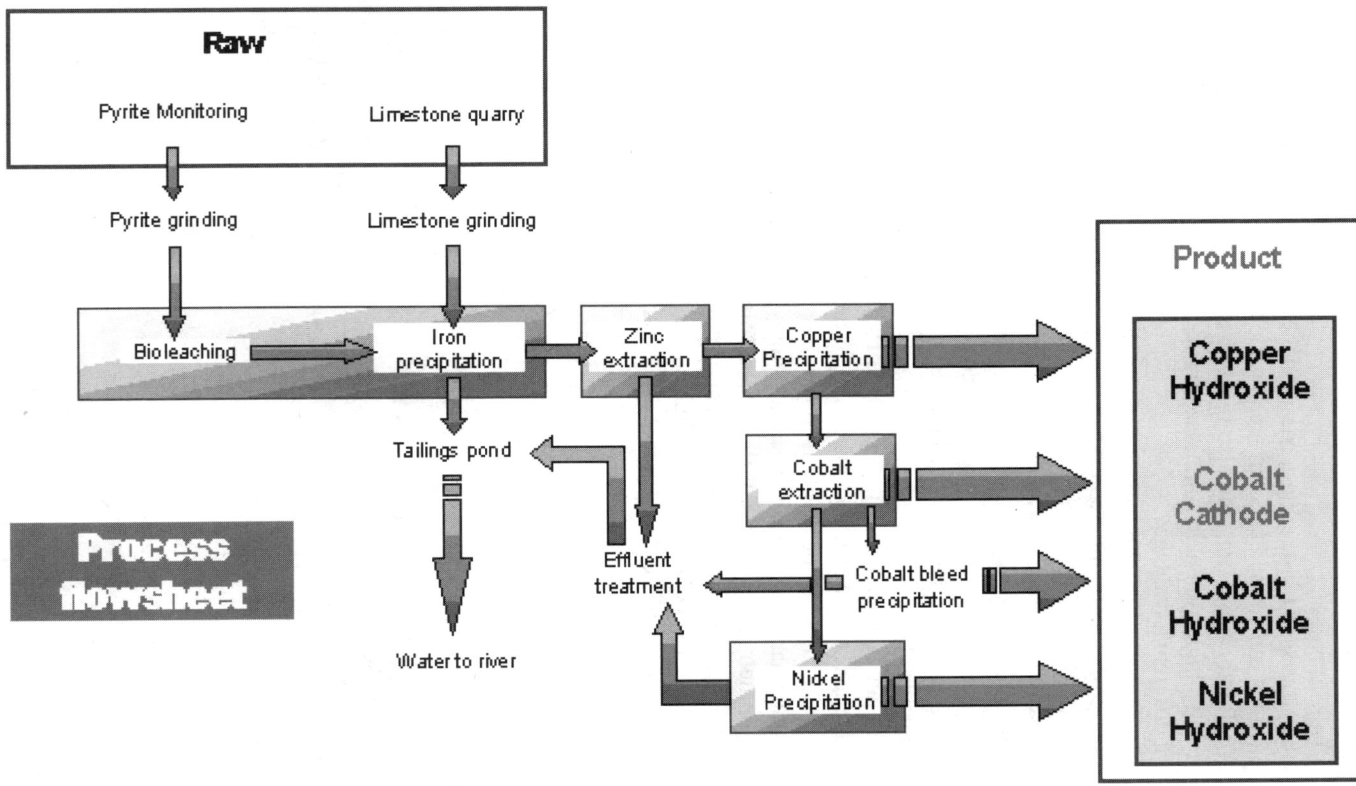

Figure 8. Simplified Ugandan bioleaching process [10]

5.4. BIO-TECHNOLOGIES

Nowadays the process of bio-leaching, based on the use of bacteria for catalyzing chemical reactions can be used at commercial scale to oxidize sulphides, mainly for the treatment of refractory gold ores, but also for the treatment of different concentrates or residues, having a high content in sulphides (pyrite, arsenopyrite, chalcopyrite, etc.). They present different advantages:

- The operations are carried out under soft conditions (30-90° C instead of 200-300° C in the case of autoclaving); consequently, the process does not require high-tech equipment or highly-qualified personnel.
- The process is environmentally safe, cost-effective and flexible.
- It offers the possibility to produce metals in remote places, on the mine site, avoiding expensive transport of the concentrates to pyrometallurgical facilities.

As an example there could be mentioned the treatment of cobaltiferous pyrite by bio-leaching in Uganda [11]:

The industrial plant has the capacity to produce 1000 tons of electrolytic cobalt (or 5% of world production) per year using bioleaching as an alternative to roasting for treating a pyritic concentrate rich in cobalt. Such concentrate results from a previous copper-mining operation unable to valorize the cobalt, finely disseminated in pyrite.

From an environmental point of view, the project solves the problem caused by the discharge trail of concentrate due to erosion and natural bio-leaching of the sulphides from the Kasese site to the Queen Elizabeth National Park.

Figure 8. presents the simplified flowsheet of the operation and figure 9. - the industrial plant.

Figure 9. The Ugandan industrial plant

6. Conclusions

The concept of Sustainable Development could reconcile the different stakeholders for the development of the Mining Industry by taking into consideration the different objectives of industrial and social development, as well as those of environment protection.

The development of the mining industry in general having a strong multiplying effect on the development of the whole industry, all possible efforts should be made at the regional scale to promote such development of the mining activity. It requires from the part of the administration to define a harmonized governmental approach at regional scale to the development of the mineral resources, based on legislation, taxation and transport systems, the economic development, environment protection rules, etc.

The attraction of national and foreign capital for the development of the Mining Industry requires an adequate legislation, a proven social, political and economical stability to make the investor confident, and also an access to the information in a modern format.

There is a good potential for the development of the Industrial minerals extractive industry at regional scale, taking into consideration the internal markets in chemical industry, petroleum industry, building industry, and the importance of transport costs for imported materials.

New tools are very powerful for improving the sustainability of the mining industry, both at the administrative and the industrial levels. They offer the possibility to make a better use of available information and to share it with the different stakeholders, to simulate and compare different development scenarios, to improve the efficiency of exploration work, to better design and conduct industrial plants in order to attain environmental and economic peak performances, and to introduce a new process, more respectful of the environment.

7. References

1. Braux, C. (1997) SynARC pour ArcView GIS. L'intégration des données "sciences de la terre" à votre décision, ESRI97, *Actes de la première conférence française des utilisateurs ESRI*, Hermes, Paris, 17-18 septembre 1997.
2. Broussaud, A., Guyot, O., McKay, J. and Hope, R. (2001) "Advanced Control of Sag and Fag mills with comprehensive or limited instrumentation", *SAG 2001,* Vancouver-Canada, Sept.30-Oct.3, 2001.
3. Chevrel, S. and Coetzee, H., (2000) A new tool for assessing the environmental impact of mining activities. Application to the surface and groundwater sensitivity analysis of the West Rand Area. Gauteng Province-South Africa, *28th International Conference on Remote Sensing of Environment*, Cape Town, March 27-31, 2000.
4. Cliff, A. (1999) *Workshop on Central & Eastern European Minerals & Metals*, Budapest, March 16th-17th, 1999.
5. Commission of the European Community (2000) *Communication from the Commission, Promoting sustainable development in the E.U. non-energy extractive industry*, May 3rd, 2000.

6. Council of the European Union (2000) Decision dated June 29th, 2000.
7. Elkington, J.(1997) *Cannibals with Forks - the triple bottom line of XXI Century Business*, Capstone Publishing, Oxford, UK, *www.sustainability.co.uk.*
8. Euromines (1999) *Guidelines on sustainable development for the European mining sector,* May 1999. See *www.euromines.be.*
9. Graves, R. (1958) *Greek Myths*, Cassel & C°, London.
10. Guillaneau, J.C., Durance, M.V., and Morin, D. (1999) Biotechnology and process simulation: two tools to develop environmentally friendly processes of ores, *Global Metals Environment – GME'99*, Beijing, China, May 1999.
11. Morin, D., Ollivier, P. and Hau, J.M. (1995) Treatment of a cobaltiferous pyritic waste by a bioleaching process. *International Symposium of Waste Processing and Recycling in Mineral and Metallurgical Industries II*, Vancouver, British Columbia, August 20-24, 1995, Proceedings, pp 23-33.
12. Owens (1994), in *Mineral Resources and Sustainable Development: A Workshop*, Elincourt, France, May 1994. BRGM document-237, 156 pages.
13. Prévôt, J.C., Dangeard, A., Martel-Jantin, B., Warhurst, A., Laurent, S., and Andriof, J. (2000) Voluntary Initiatives and Sustainable Development in the Non-Energy Extractive Industry, *Final Report, European Commission - Enterprise Directorate-General,* 65 pages.

MINERAL RESOURCE BASE OF GEORGIA IN THE XXI CENTURY

A. G. TVALCHRELIDZE
A. Janelidze Geological Institute, Georgian Academy of Sciences
1/9 M. Alexidze St., 380093 Tbilisi, Georgia
E-mail:_ sandro@kheta.ge

1. Introduction

Our infantile euphoria caused by destruction of the Berlin Wall is entirely forgotten nowadays. The whole geopolitical space, recently known as the *"communist block"*, is experiencing a severe economic and political crisis. We commence to understand only now that any serious change in political and economic systems is impossible without irreversible, incongruent alteration of the social structure of the society and social conscience of population, and that consequences of those alterations are often unforeseen.

By the end of the XX century the Newly Independent States found themselves in the lowest phase of a severe structural crisis. The rest of the World, and, first of all, developed countries, are unable to ignore this circumstance (restore the iron curtain). Belonging to the same geopolitical space and having the common economic history during the XIX and XX centuries, the CIS countries will be able to overcome this crisis and to integrate into the world community only step-by-step, exploiting their resource bases and developing competitive branches of industries. The article hereto is the first publication, dealing with the problem of the mineral resources of Georgia and modern methods of its economic evaluation and appraisal.

2. Macroeconomic Definition of the Resource Base

The short and tragic history of our independent statehood reveals a simple but important fact: after the collapse of the USSR and first steps in free-market economic relations of the Newly Independent States, macroeconomic approach to the resource base and its governmental management has to be reviewed. According to the classical definition (see, e.g. [12]), the mineral resource base implies a totality of mineral resources of a given territory, which may be exploited or used otherwise under stated economic conditions and currently feasible technology or technology which will be feasible in the near future.

In the Soviet Union the resource base was always considered to be a political-economic category, helping the enormous country to survive and develop independently from the

A.G. Tvalchrelidze and G. Morizot (eds.),
Mineral Resource Base of the Southern Caucasus and Systems for its Management in the XXI Century, 19–70.

civilized world, beyond the rigid iron curtain. In other words, the USSR had to be provided by any resource when needed, and the resource had to be of the local origin. Thus, nobody was surprised, when uranium ores derived from Ukrainian or Middle East mines were processed and treated at the Uralian secret plants: state security reasons were decisive, and nobody paid any attention to the economic imperatives. Due to such framework, any economic branch was managed from the single centre – Moscow, whereas prices on natural resources and products of their processing played a role of stocktaking units. Any relation between different industries was effectuated on the governmental level only. Such "managerial chart" had created a rigid, vertical, pyramidal infrastructure of each branch with minimum "horizontal" relations between different industries or enterprises. Thus, any enterprise in any part of the immense country was supplied with materials and sold its production via Moscow ministerial offices.

On the other hand, according to the Anglo-Saxon macroeconomic tradition [12, 30], the resource base must be considered a specific form of the tangible property that takes part in the world market processes and determines origination of the added value. Thus, the resource base or any of its parts may be sold, alienated, leased, landed, mortgaged, taken as warranty, given for exploitation, according to existing legislation, to any physical or legal person, etc.

Table I contains basic differences in philosophical and methodological approaches to the mineral resources that existed in the Soviet Union and the rest of the civilized world.

TABLE I. Soviet and Western Approaches to Mineral Resources

Criterion	Soviet Approach	Western Approach
Discipline	Geology of mineral deposits	Economic geology
Subject of investigation	Earth's bowel	Mineral resource
Geological body	Mineral deposit	Mineral deposit
Evaluation principle	Industrial	Economic
Principles of reserve calculation indices	Geological conditions[1]	Ultimate profitability

The last row of the table needs some additional explanations. The geological conditions were approved by the centralized State Commission on Reserves for the USSR in whole. Table II provides information on the framework existing before the collapse of the Soviet Union.

---

[1] In the Soviet geological literature geological conditions implied minimum contents of the economic constituent and its minimum marginal reserves, above which industrial exploitation of the latter became technically and economically feasible (see, for instance, [39]).

TABLE II. Geological Conditions for Metallic Deposits (after [39])

Category	Metals	Minimum reserves	Minimum content, %
Ferrous metals	Fe, Mn	nx100,000 t	20-25
Non-ferrous metals	Cu, Pb, Zn, Ni	nx10,100 t	0.4-1
Rare metals	W, Mo, Sn, Hg, Li, Be	nx1,000 t	0.1-0.2
Radioactive metals	U, Th	nx1,000 t	0.05-0.1
Noble metals	Au, Re, Pt, Ru	Kilogrammes	0.0005

The table above reveals some very interesting features:

- Geological conditions of metals in deposits are directly proportional to their clarks in the Earth's crust – the higher the clark is, the higher geological conditions are.
- Market prices of metals in all cases are in reverse logarithmic proportionality to their clarks in the Earth's crust.
- Radioactive metals have the same distribution peculiarities as rare elements (and geochemically they belong to this category) but their minimum content is twice less than those of rare metals. This feature may be explained by the extremely important role of the latter for the Soviet militarized industry.

Again, technical progress and, namely, improvement in metal enrichment technologies created methods, which made feasible treatment of ores with gradually decreasing contents of economic constituent. Figure 1 demonstrates, for instance, mean gold content in ores processed in different years throughout the world. Two bending of the curve is observed. The first one is related to introduction of the flotation methods of ore enrichment, whereas the second one is due to introduction of the cyanide treatment and, partially, heap leaching. Thus, simple decreasing of minimum content in books of the State Commission on Reserves determined automatic increase of the balance reserves of the whole country (see Figure 2).

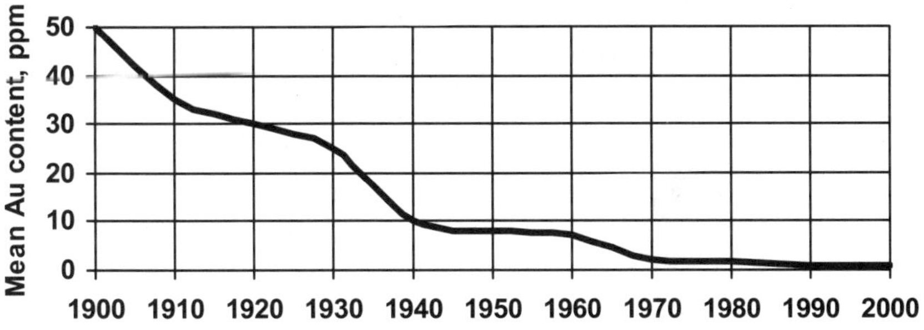

Figure 1. Mean Gold Content in Active Mines in Different Years

In the Western countries reserves of mineral resources are treated using a quite different approach. The whole mineralized space is divided into a number of elementary blocks approximately corresponding to 1 t or 1 m^3 (depending on the type of a mineral resource). In each block total price of the leachable constituents is calculated. If this number should ultimately exceed exploitation cost, amount of ores (and metals) in the block is added to the deposit reserves. Appropriated methods of computer modelling made this technology feasible even for a single analyst.

Frankly speaking, this theory is simple enough but, unfortunately, its corollaries have never been analyzed seriously. Hence, they lead to a total semantic misunderstanding between Soviet and Western specialists because even the simplest notions such as, for instance, "metal reserves" are understood, as shown, in a quite different way. Such misunderstanding appeared for the first time in the course of huge Western investments in the Soviet resource base. That is why I tried to elaborate some methods, which are presenting achievements of the Soviet metallogenic science in the terms, acceptable for the Anglo-Saxon macroeconomic tradition.

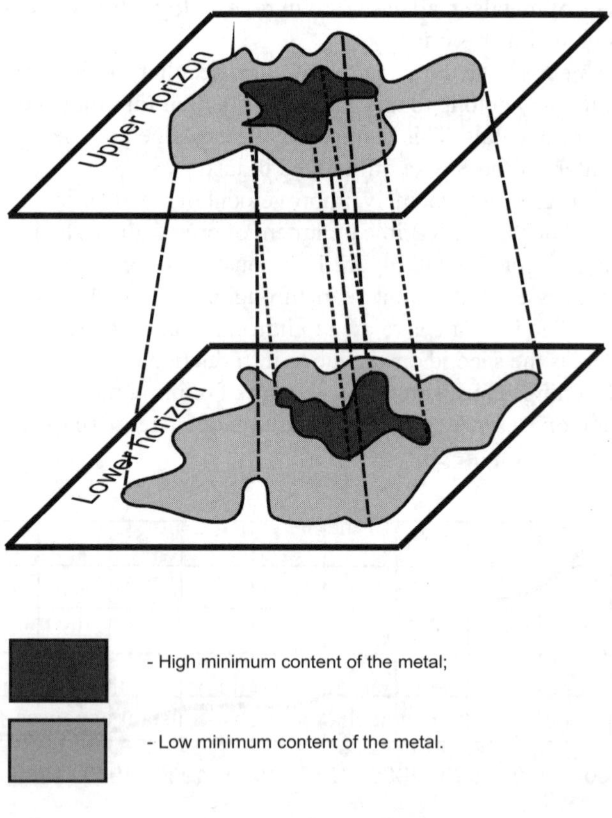

Figure 2. Ore body Contours in Conditions of Different Minimum Contents of Metals

3. Macroeconomic Methods for Evaluation of the Resource Base

A very simple index has been proposed: a "Price Index" of any mineral resource or of the entire resource base. The latter is referred to as an ultimate income from overall exploitation of a mineral resource or the resource base (expressed in long-term market prices) minus all exploitation, transport, industrial, prospecting and exploration [49]. Then, for non-renewable resources:

$$I_p = P_S Q_O, \tag{1}$$

Where: I_P = price index; P_S = specific price of 1 metric unit (ton or m^3) of a mineral resource; Q_O = reserves of a resource. For renewable resources equation (1) should be changed by:

$$I_p = P_S Q_S t, \tag{1a}$$

Where Q_S = exploitable annual reserve of the resource; t = term of depreciation of basic assets including equipment and engineering.

Specific price of natural resources depends on a lot of factors including ratio between demand and supply of the resource to the world markets, economic type of the resource, characteristic features of its natural manifestation, etc. Thus, a number of indices should be measured including concentrations of economic constituents, technologic possibilities of their extraction or leaching, actual or future conditions of mining, geological, hydrogeological, engineering geological environment, physical and mechanic properties of country rocks, and the cost of the whole technologic cycle (industrial, exploitation, exploration, enrichment, maintaining of the tailing depository, stocking, transportation and other expenses):

$$P_S = \sum_i P_i C_i k_{Ei} - K_O, \tag{2}$$

Where P_i = long-term market price of the ore constituent i; C_i = its concentration in ores; k_{Ei} = coefficient of leaching of the constituent i; K_O = cost of 1 metric ton of ores including all above-mentioned expenses. Calculation of K_O is performed in the course of feasibility studies as it greatly depends on power and output of the mining enterprise. Figure 3 describes, for instance, mining and processing cost of 1 metric ton of auriferous ores at Sakdrisi deposit, Southern Georgia. It could be seen that the specific cost is divided into two categories. Under such conditions, constant expenses shall imply the totality of organization and exploration[2] expenses per one metric t of ores, whereas

[2] As mineral resources should be ultimately recoverable, e.g. the exploration expenses should be covered by the price of mined ores.

variable expenses shall imply the totality of expenses on mining (including construction of the mine and excavation equipment purchasing), flotation (including construction of the plant), gold leaching (including construction of the cyanurization factory), gold production (gold ingots), maintaining of tailing (including depository construction), transportation (including trucks purchasing), recultivation of the land plot (after actual cessation of mining) spending. It could be seen, that the more powerful the production is, the more profitable is the mining business. At the same time, output of the mining enterprise should be selected in such a manner that the term of mining corresponds to depreciation duration of main assets and equipment.

However, the overall macroeconomic analysis of the resource base may use more general regularities based on the profitability coefficient. For instance, from the very beginning the possible profitability of eventual mining enterprise is known due to the estimated tonnage/grade ratio (Singer and De Yung, 1980 [38]). This coefficient implies ratio of the price-cost difference of 1 metric unit of the natural resource versus the cost. Then, the equation (2) may be written as:

$$P_S = p_D \sum_i P_i C_i k_{Ei} , \tag{3}$$

Where p_D is a profitability coefficient.

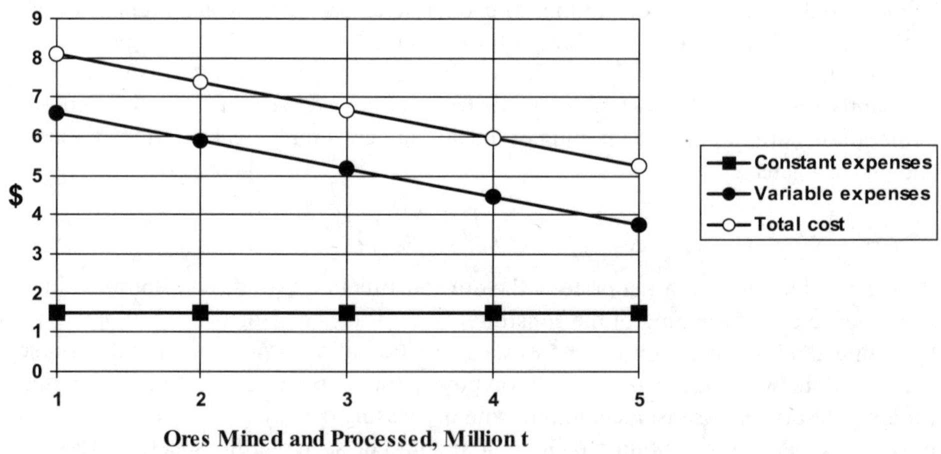

Figure 3. Cost of 1 t of Ore Processing at the Sakdrisi Deposit, Southern Georgia

4. Metallogenic Framework

From the beginning of our century and, namely, in the thirties-sixties the Caucasian region in general, and Georgia in particular, were subject to numerous geological, geotectonic, petrologic and metallogenic investigations. These materials have been thoroughly systematized in collective monographs [19, 20], where the geological structure of Georgia is considered to be based on the older geosynclinal theory. Introduction of ideas of the New Plate Tectonics presented a new challenge for local geologists [1-4]. Here the Phanerozoic history of the Caucasus is interpreted as gradational development of an active continental margin of the oceans, consequently, Palaeo- and Mezotethys.

Furthermore, an interesting approach was developed by Adamia and his collaborators [5, 6] and a number of publications by Gamkrelidze [16] who used a terrane analysis for explaining the complex structure of the Caucasus. According to the classical definition [20], a terrane (or "a tectonic-stratigraphic exotic bloc or complex") represents a geological body of the regional extent, limited by deep faults, which differs from neighbouring bodies by its geological structure and, therefore, by the geological history. According to Gamkrelidze [16], the Eastern Mediterranean mobile belt consists of five first order terranes (Figure 4). During the Palaeozoic, Mesozoic and Early Cainozoic time these terranes underwent horizontal displacements in different directions within the oceans, consequently, Proto-, Palaeo-, Mezo- and Neotethys. In Middle Cainozoic time they were gradually glued to the Euro-Asian Plate. Later, during the Late Alpine orogenesis (Neogene-Pliocene), these terranes underwent intense deformations. Thus, during the whole geological history of the Caucasus, its northern margin was active, whereas the southern one remained passive.

Figure 5 represents a sketch of tectonic and, correspondingly, metallogenic zoning of Georgia. Three first order terranes are fixed here. The Greater Caucasus terrane accreted to the Euro-Asian Plate within the Hercinian time [16]. The Baibut-Sevanian terrane has its ophiolitic suture southward from Georgia. Within the boundaries of the country it is presented by the Somkhito-Karabakh subterrane – a typical andesitic island arc of a global extent [57]. Different subterranes of the Black-Sea-Transcaucasian terrane represent different parts of the oceans Mezo- and Neotethys. Two main tectonic elements played a decisive role in lithogenesis: pelagic oceanic Chkhalta-Tphani subterrane and the Transcaucasian Median Mass, different parts of which have undergone relatively independent development [59]. Exposures of the Palaeozoic crustal basement of this median mass are fixed within the Dzirula, the Khrami and the Loki massifs.

The Greater Caucasus terrane is mainly composed of pre-Cambrian and Palaeozoic rocks. According to numerous investigations (see, for instance, [1-2, 10] and many others), within the Greater Caucasus a granitic-metamorphic complex (different crystalline schists and paragneisses) has been intruded by Palaeozoic granites. During the Hercinian tectonic events this terrane was accreted to the Euro-Asian Plate [5, 16]. Shallow sea sandy-argilleous rhythmical sediments with lenses of limestones present

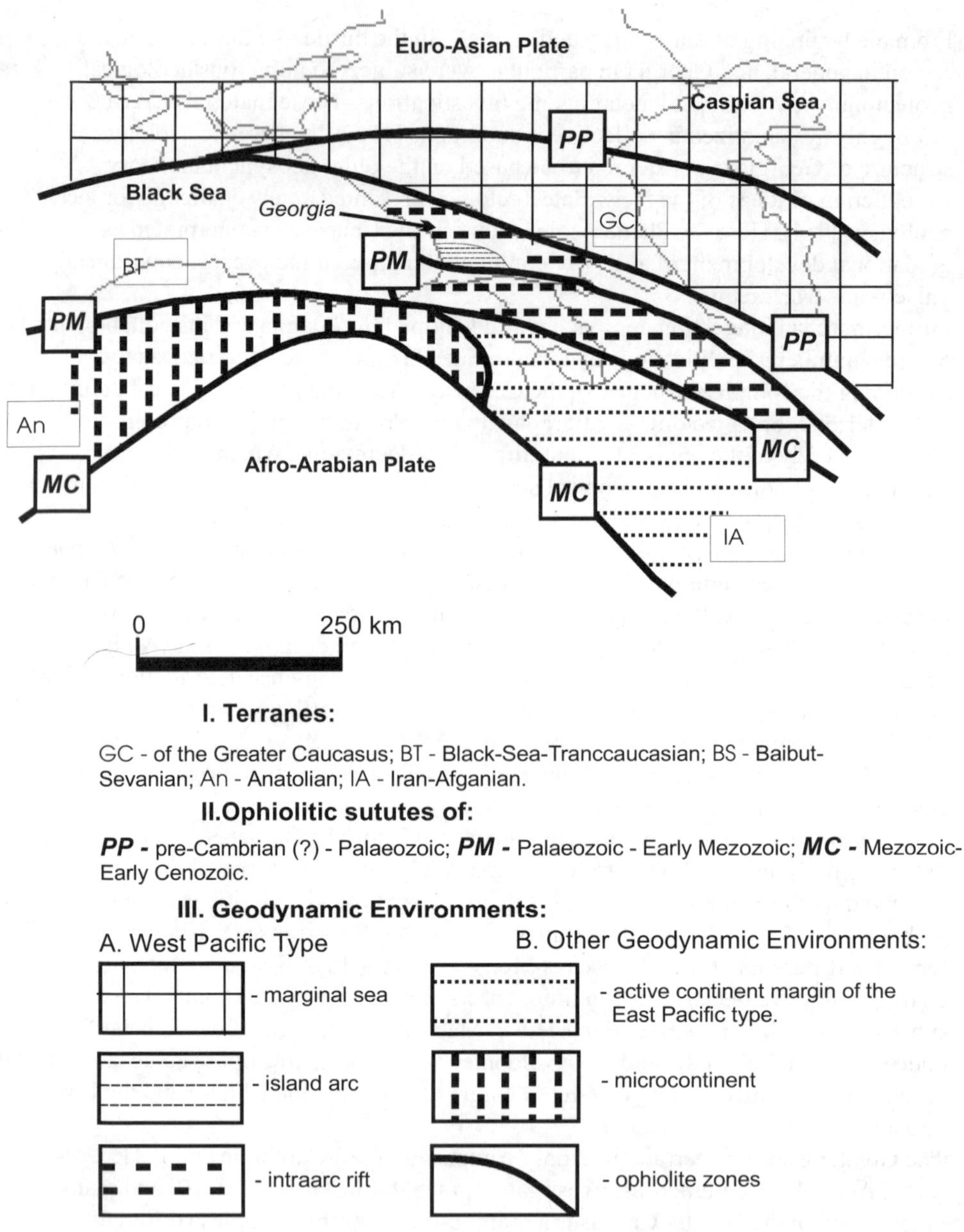

Figure 4. Tectonic Setting of the Eastern Mediterranean [16]

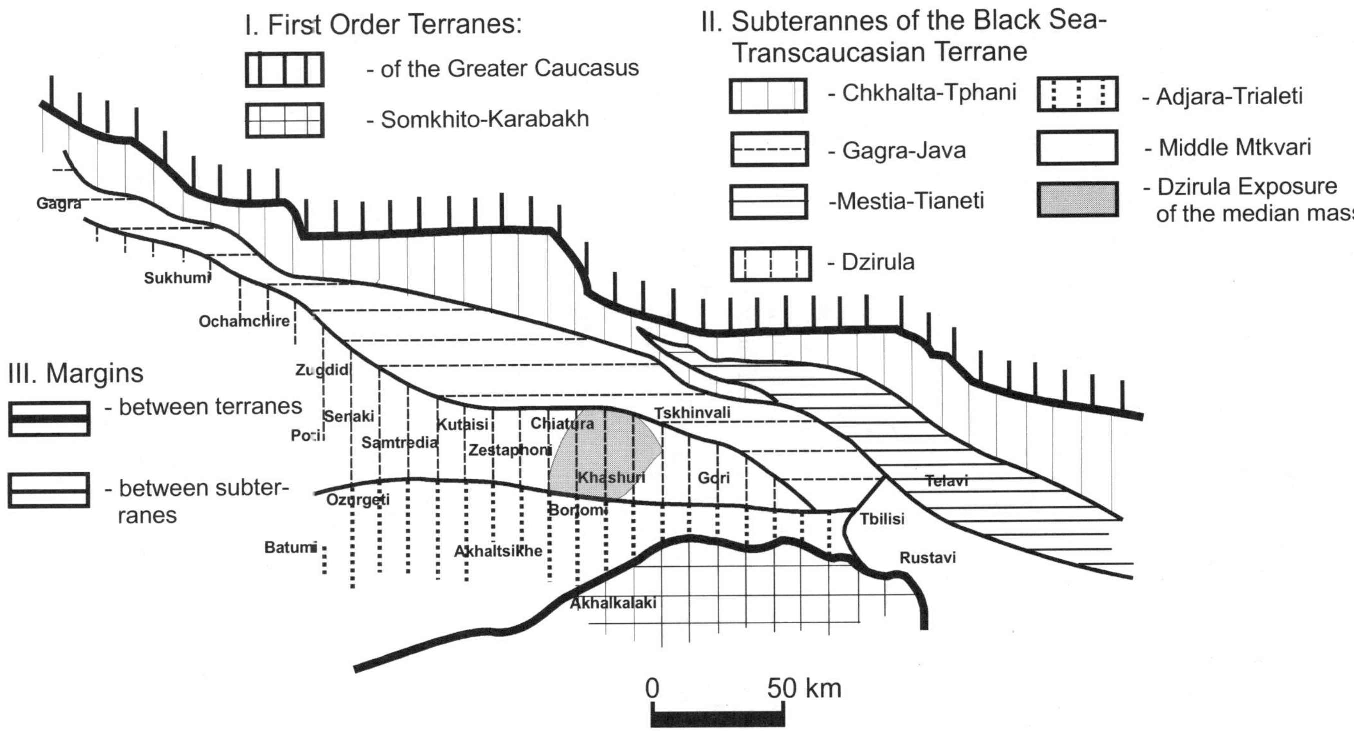

Figure 5. Tectonic and Metallogenic Zoning of Georgia [16

lower Jurassic rocks here.
The Black-Sea-Transcaucasian terrane is separated from the Greater Caucasus terrane by the Greater Caucasus Main Fault that played a role of a subduction zone in Early Jurassic [1-5] and after the Bathonian (Middle Jurassic) orogenesis was modified into a typical deep thrust. The Dzirula outcrop of the Transcaucasian Median mass occupies the central part of the Black-Sea-Transcaucasian terrane. Presence of pre-Cambrian (?) ophiolites has been announced here [40]. This massif consists of Palaeozoic schists intruded by numerous Palaeozoic and Jurassic gabbros, quartz diorites, granites accompanied by pegmatite veins. Palaeozoic rocks are overlapped by shallow water Jurassic clays and Cretaceous limestones. By that time Dzirula Massif lost its independence and Mezozoic rocks were concordantly accumulated into the Dzirula and the Middle Mtkvari subterranes.
The main sequence of the Chkhalta-Tphani subterrane consists of a thick (up to 3-10 km in different places [17]) Lower Jurassic series of black schists (argillitic deep sea sediments with organic material metamorphosed up to green schists) enriched at different levels by metamorphosed limestone strata and fissure flows of tholeiitic bimodal basalt-rhyolitic volcanism [9, 10, 42]. Early Jurassic deep-sea sedimentation ended in the Aalene time, and the Chkhalta-Tphani subterrane took no part in the Alpine tectonogenesis.
Middle Jurassic (Bajocian) basalt-andesite-dacite-rhyolite formation is typical of the Gagra-Java subterrane. In the beginning of the Middle Jurassic this structure was broken out of the Transcaucasian Median Mass and was involved in the Mesotethyan events [15]. The lower part of the thick (up to 3 km) volcanic sequence is composed of tholeiitic fissure pillow lavas. In Middle Bajocian the calc-alkaline andesitic central-type volcanoes were originated [13, 26], and the Gagra-Java trough was consequently transformed into an island arc [31]. In Bathonian time carboniferous molassic sediments were deposited here. Simultaneously, granitic stocks were intruded just in the marginal suture between the Gagra-Java and the Chkhalta-Tphani subterranes. Jurassic limestones (including reef) lie at the top of the stratigraphic column.
The Mestia-Tianeti subterrane originated in the Middle Jurassic time. The stratigraphic sequence is presented by calcareous sandstones and limestones followed by the flysh sediments with rare lenses of limestones.
In the Dzirula and the Middle Mtkvari subterranes Jurassic rocks are presented by the carbonate sequence. In the Dzirula subterrane the thickness of Upper Jurassic limestones gradationally decreases towards the Dzirula Massif, forming several subsequent reefs near Kutaisi city.
Before Cretaceous all described subterranes of the Black-Sea-Transcaucasian terrane were already amalgamated, and Cretaceous carbonate sedimentation took place in centripetal succession from the Dzirula Massif. Thus, shallow-water organogenic limestones are typical of the southern part of the Gagra-Java and Mestia-Tianeti subterranes. Within the Dzirula and the Middle Mtkvari subterranes carbonate sediments of this age are relatively thicker (up to 500-700 m) and often interlaid with andesitic lavas and pyroclastics.
Cainozoic sediments are practically absent in the Gagra-Java and the Mestia-Tianeti

subterranes. Only thin (50-150 m) Palaeogene shallow-sea carbonate and flyshoid sequences are presented here. In the Dzirula and the Middle Mtkvari subterranes Palaeogene sediments lie in the basement of, correspondingly, the Rioni and the Mtkvari river troughs, and their thickness gradually increases towards the Black and the Caspian seas. Palaeogene sediments are followed by the Oligocene Maikop series - an intermountain molassic formation with clayey, often rhythmical rocks and few limestones. This series underwent intense dia- and autigenesis [29]. Thickness of this series increases westward and eastward from the Dzirula Massif. In the central parts of the Rioni and the Mtkvari troughs the Maikop series hosts hydrocarbon occurrences whereas in vicinity of the Dzirula massif rich and large manganese deposits were formed (the Chiatura Mine). Neogene sandy-clayey sediments in the Dzirula and the Middle Mtkvari subterranes form a thick (up to 2 km) sequence. Near the Black Sea cost Neogene rocks are superimposed by shallow marine Quaternary sandy-clayey sediments of the consequent littoral terraces.

In Early Cainozoic, in the southern part of the Black-Sea-Transcaucasian terrane a rift trough of the Adjara-Trialeti subterrane started to develop directly upon carbonate Cretaceous sediments (see, for instance, [2, 5, 26], etc.). Activity of Palaeogene and Early Neogene trachibasalt-trachiandesite volcanism increased from the eastern edge of this rift towards the west. In Late Neogene a number of quartz-monzonite-diorite massifs were intruded predominantly in the western, tectonically more active, part of the said rift. Orogenesis occurred here by the end of Neogene, just before the Quaternary period.

In the Somkhito-Karabakh terrane[3] Middle Jurassic rocks lie directly upon the crystalline basement of the Transcaucasian Median Mass. The sequence starts with a basal conglomerate that overlaps Palaeozoic granites, for instance, near the Khrami and the Loki massifs, followed by arkose sandstones. Outside Georgia, in Armenia and Karabakh, a thick calc-alkaline subaquaous andesitic volcanic-sedimentary series is developed [37]. But northward, towards Georgia, the thickness of this island arc affinity gradually decreases, and the manifestations of Middle Jurassic volcanism become rare [27]. Here a thick ore-bearing (Madneuli Mine) Cretaceous andesitic (minor basalt-andesite-dacite-rhyolite) calc-alkaline formation has created numerous volcanoes and vast fields of acid subaqual pyroclastics in intervolcanic areas. This sequence is especially thick (up to 1-1.5 km) within the Bolnisi Mining District [27]. Palaeogene and Neogene strata continue the Cretaceous volcanic sequence forming andesitic subaerial affinity of the subsequent calc-alkaline series. The most interesting event is presented by thick (several hundred m) flows of massive andesitic aerial lavas that form a volcanic plateau of the Southern Georgia. The same phenomena (but much less intense) have been encountered on the southern slope of the Greater Caucasus [37].

The modern *premier case* metallogenic investigation was effectuated by my father, Academician George A. Tvalchrelidze, who published in 1961 an important monograph

---

[3] In reality this unit represents a subterrane of the Baibut-Sevanian terrane. On the Georgian territory this subterrane is the only one. Due to this fact we will use later the term "Somkhito-Karabakh terrane" for distinguishing it from subterranes of the Black-Sea-Transcaucasian terrane.

[54]. This publication explained metallogenic features of Georgia from eugeosynclinal point of view. Later he extended his approach entirely on the Mediterranean mobile belt [56]. Introduction of the New Plate Tectonics obliged the author to change his metallogenic method. First of all, he paid attention to typomorphic metallogeny of global volcanic belts [57] and then, in early 1980s published the first "mobilistic" metallogenic schemes of Caucasus [58, 59].

Nowadays there is a possibility to interpret main metallogenic features of Georgia basing on the above-mentioned terrane analysis. Table III describes typical Georgian ore/mineral formations of different mineralization epochs in different terranes (subterranes).

G. Tvalchrelidze [55] distinguished the Proterozoic, Palaeozoic, Caledonian, Hercinian, Cimmerian (early-Middle Jurassic), and Alpine metallogenic epochs of Caucasus. The first two of them have been manifested within the Northern Caucasus pre-Cambrian-Palaeozoic ophiolite belt and the related volcanic island arc: the Tyrny-Auz-Pshekysh suture and the Caucasus Foreridge [2, 6]. In Georgia a single manifestation of pre-Cambrian – Early Palaeozoic mineralization is observed within the Dzirula Massif (Rikoti deposit of facing gabbros), where the exact age of the formation is almost impossible to establish. We were obliged, therefore, to distinguish a single pre-Hercinian epoch without its exact dating.

Cimmerian (Early Jurassic) mineralization in different subterranes is manifested in deep-ocean (Cu-pyrrhotite ores and roof shales in the Chkhalta-Tphani subterrane) and shallow-water marine (refractory clays in the Dzirula subterrane) environments.

George Tvalchrelidze [54, 55] refered a single Alpine epoch, which occurred in different metallogenic zones to a slightly different time. But further thorough observations on ages of almost all mineralizations clearly indicate presence of, at least, three different epochs in Cainozoic. The first of them occurred in Late Cretaceous, the second one took place in Oligocene time, and the third epoch manifested itself in Neogene. All of them are characterized by different mineral formations in different terranes and subterranes.

Early Alpine epoch produced two typical island arcs at the northern and southern peripheries of the Transcaucasian Median Mass. The former (Gagra-Java subterrane) is characterized by vein and strata-bound Pb-Zn and barite deposits as well as iron-bearing coal basins. The latter (Somkhito-Karabakh terrane) produced auriferous polymetallic base metals, facing tuffs, lithographic stones. The central part of the Black-Sea-Transcaucasian terrane by that time was characterized by orogenic environment producing facing materials and shallow-water Al-bearing analcite sandstones which have been deposited in a closed basin of the Dzirula subterrane. In the shallow-water basin (Mestia-Tianeti subterrane), superimposed upon a newly consolidated continental margin (Chkhalta-Tphani subterrane), auriferous realgar-orepigment mineralization is been formed.

Middle Alpine epoch is manifested only in the central part of the Transcaucasian Median Mass. During the Oligocene a transgression westward and eastward from the Dzirula Massif shallow-water molassic basins was originated. Molassic sediments, inclined towards, consequently, the Black (Dzirula subterrane) and the Caspian (Middle Mtkvari subterrane) seas, were subject to specific sedimentation with typomorphic dyagenetic

TABLE III. Main Ore/Mineral Formations of Georgia

Main Terranes (Subterranes)	Mineralization Epoch						
	Pre-Cambrian and Early Palaeozoic	Hercinian	Cimmerian (Early Jurassic)	Early Alpine	Middle Alpine	Late Alpine	Modern
Greater Caucasus						Mo, As,W, Sb	
Chkhalta-Tphani		marbles marmorized	Cu, roof shales, facing limestones			Hg	Au placer
Gagra-Java				coal, Fe, Pb-Zn, barite, facing, building and flux limestones, anhydrite, gypsum, lime	phosphorites	cement and building lime-stones, anhy-drite, gypsum, foundry sands	foundry sands, brick earth, mineral paints
Dzirula	facing gabbros	talc, pottery pegmatites, facing granites and quartz diorites	Facing marmorized limestones, refractory clays	Al, gumbrine, porcelain kaolinite, facing teschenites, facing marmorized and building limestones, lime, dolomites, chalk	oil, thermal waters, Mn, chalcedony	gumbrine, quartz sands, lime, cement limestones	peat, mineral paints, cement clays, brick earth
Mestia-Tianeti				Au-As, facing marmorized, cement, building limestones, lime			foundry sands, brick earth
Ajara-Trialeti					thermal waters, Fe, quartz sands	coal, Au-Pb-Zn-Cu, zeolites, diatomites, facing syenites, marmo-rized limestones, gypsum,refractory clays, agate, opal	peat, acid-resisting andesites, facing basalts, cement clays
Middle Mtkvari					oil, gas, halloides	flux limestones	Glauber salt, gravel, foundry sand, cement clays
Somkhito-Karabakh				Fe, Au-Ag-Cu-Pb-Zn-barite, lithographic stones, flux limestones, facing tuffs		perlite, agate, obsidian	Au placer, foundry sands, facing basalts, cement scoriae anhydrite

and autigene mineral formations. Manganese deposits, manganese-related spongolite and chalcedony mineralizations are typical of this epoch. Simultaneously, in lowland areas, where molassic sediments were thicker, first accumulations of hydrocarbons occurred, locally associating with haloids.
Late Alpine epoch was complex. Within the Greater Caucasus terrane and along its Main Fault typical orogenic, both granite-related and telethermal, ore formations were originated. Southward, in shallow water conditions (Gagra-Java subterrane) phosphorite ores were deposited. In the central part of the practically entirely consolidated Black-Sea-Transcaucasian terrane shallow-water quartz sands were deposited. In the Adjara-Trialeti subterrane a typical porphyry copper system as well as iron skarn deposits were related to quartz-monzonite-diorite intrusive bodies. Simultaneous postvolcanic processes resulted in agate formation. In shallow subsequent fresh-water lakes and bogs, correspondingly, diatomites and carboniferous strata were deposited. In the Somkhito-Karabakh terrane last volcanic paroxysms and a postvolcanic activity produced perlite, obsidian and agate deposits.
Quaternary epoch was governed by climatic conditions of the Greater and the Lesser Caucasus as well as by features of their relief. Different alluvial (including Au placers), diluvial and proluvial sediments, weathering core and modern bog environment determined origination of corresponding modern mineral deposits.

5. General Characteristics of Georgian Mineral Resources

Among Georgian mineral resources fuel, geothermal sources, ferrous, precious, non-ferrous, light, rare metals and elements, chemical and agrochemical, inert, facing and building materials, as well as mineral waters may be distinguished. Appendix contains generalized information about reserves of the most important mineral deposits in Georgia.

5.1. FUEL AND ENERGY MATERIALS*

Several oil and a single gas fields of Georgia are distributed in both the Middle Mtkvari and the Dzirula subterranes, the former continuing the Azerbaijan oil fields and the latter belonging to the Black Sea basin, the marine part of which has not been evaluated yet. These occurrences are related to the Middle Alpine epoch of mineralization, when primary accumulations of hydrocarbons took place [20]. After the Neogene transgression these hydrocarbons migrated to dome structures forming oil and gas fields. According to the recent investigations, bulk reserves of crude oil of these fields are calculated to be 33,000,000 tons and economic resources - 92,000,000 tons (excluding the Black Sea marine zone within the limits of the Georgian exclusive economic zone). Petroleum is of a high quality and belongs to the naphta type [20]. Its density varies

---

* See in Figure 6.

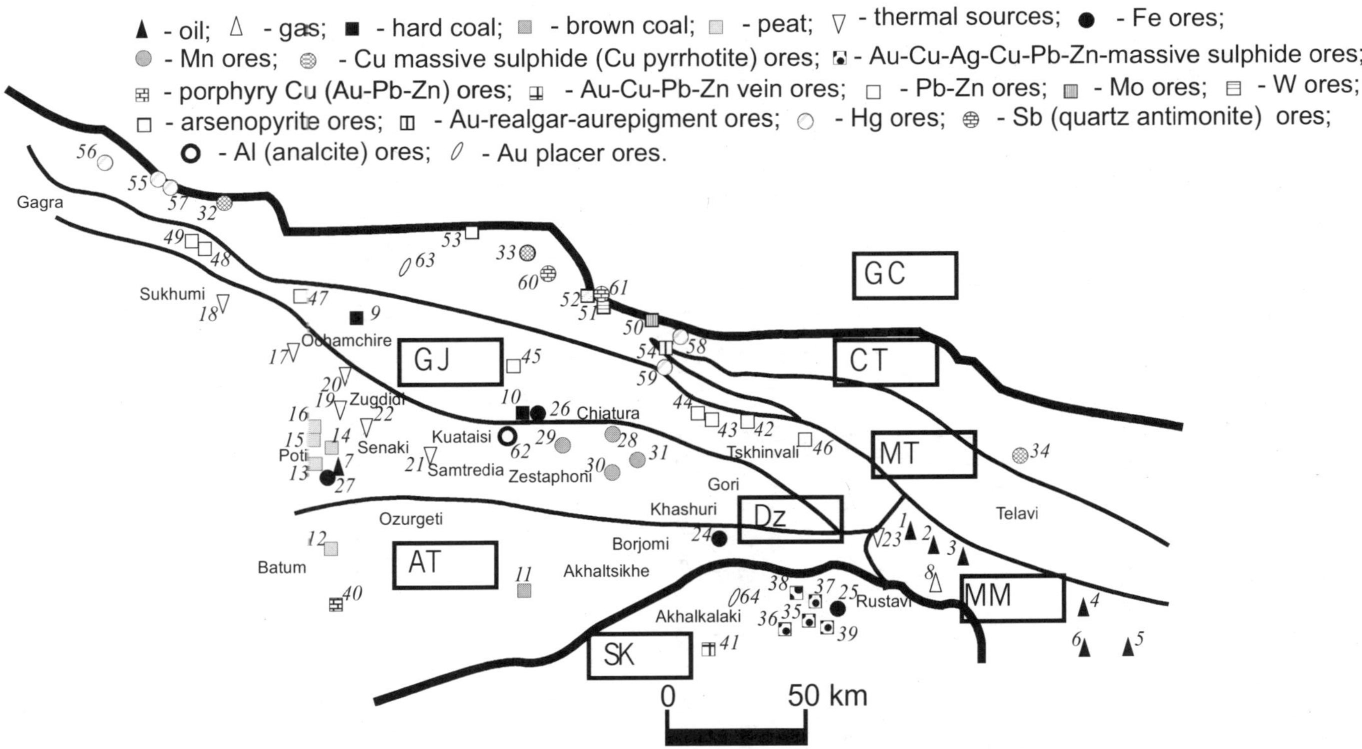

Figure 6. Distribution of Energetic and Metallic Deposits in Georgia

between 0.856 and 0.878 g/cm^3; mean quantity of oils is 42.7% and of those of the light fraction - 38.23%. Petroleum is characterized by a low percentage of sulphur.
Georgian coal deposits are related to two mineralization epochs. During the Early Alpine time hard coal deposits were connected with the carboniferous molassic Middle Jurassic sequence in the Gagra-Java subterrane. These sediments underwent intense dyagenetic and autigene alterations during Early and Late Cretaceous sedimentations. In Late Alpine epoch the Akhaltsikhe brown coal mine (Adjara-Trialeti subterrane) was due to reducing sedimentation in lacustrine and bog conditions that happened between volcanic paroxysms. The Tkibuli-Shaori and the Tkvarcheli deposits include 347,000,000 tons of hard coal whereas the Akhaltsikhe deposit hosts 71,000,000 tons of lignite. Deposits of hard coal lie in the Bathonian molassic basins and form several horizons. Coals belong to humus and leptobiolite types. Their calorific capacity varies from 4,100 up to 5,600 ccal/kg. Ash content is equal to 1.5-8% in average. The Akhaltsikhe lignite deposit is due to Lower Neogene sandy carboniferous sequence among the shallow water sandy-clayey sediments followed by coarse-grained tuffaceous rocks. Calorific capacity of coal in different horizons varies from 3,230 up to 4,500 ccal/kg, ash content – from 13.03 up to 45.00%.
Among the 46 peat manifestations of Georgia five have industrial importance. Bulk peat reserves of these bogs exceed 71,000,000 m^3. All of them are related to the Kolkheti Lowland and are distributed along the Black Sea beach. Peat belongs to the lowland type, mean ash content varies from 25 up to 45%, and natural moisture is equal to 88-91%. Peat is characterized by low acidity (pH = 5-6) and N content of 1.3-1.5% [20].

5.2. GEOTHERMAL SOURCES*

Numerous thermal sources and vents represent important geothermal resources of Georgia; few of them have been drilled. Energetical capacity of these drills is calculated to be equal to 2 billion kWh equivalent. Two main basins of thermal sources may be distinguished. In the Kolkheti Basin (Ohurei, Kindgi, Zugdidi, etc.) thermal waters have low mineralization degree; they are overheated (103-105° centigrade at vents) and characterized by high static pressure (13-17 atm). In the Eastern Georgia Basin (Lisi drill near Tbilisi) mineral waters have high sulphur and methane contents and temperature of about 65°C at the vent. The first group of thermal waters may be used for electricity production, hot water supply, and civil heating, whereas the second group is applicable for civil heating only and is used in balneology.

5.3. FERROUS METALS*

Georgia does not belong to important iron-bearing provinces. At the same time, bulk reserves of Fe_2O_3 of its 4 commercial deposits (Tkibuli-Shaori, Dzama, Poladauri and Supsa-Natanebi) are equal to 226,000,000 tons. Reserves of each deposit are given in

---

* See in Figure 6.

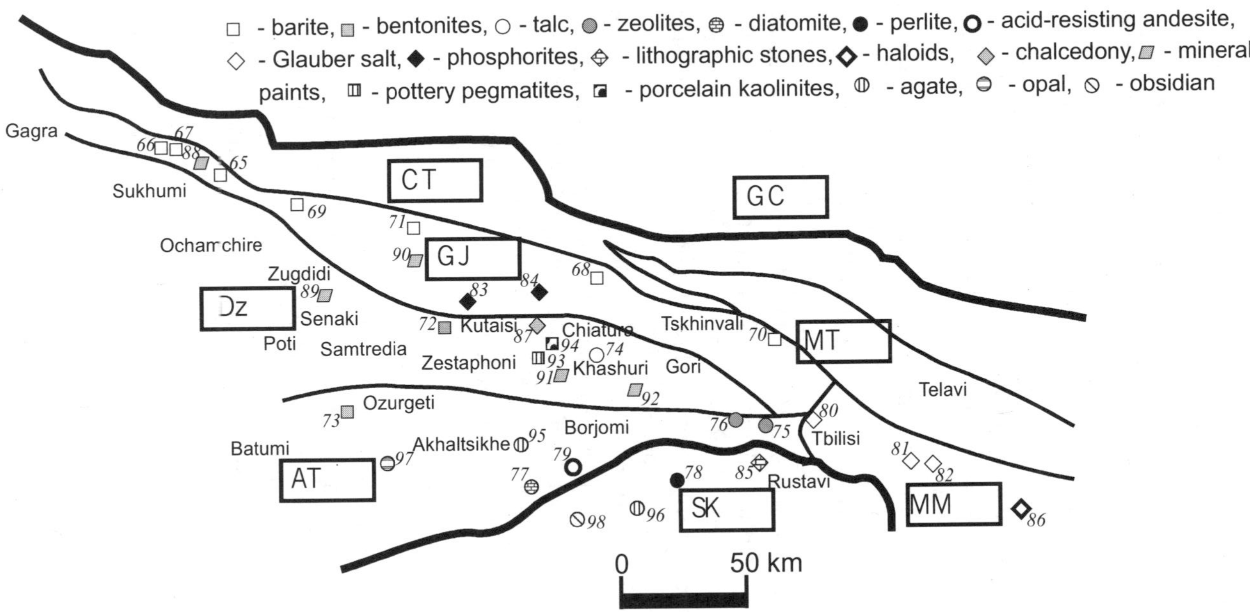

Terranes and subterranes: GC - of the Greater Caucasus, CT - Chkhalta-Tphani, MT - Mestia-Tianeti, GJ - Gagra-Java, Dz - Dzirula, AT - Adjara-Trialeti, SK - Somkhito-Kakabakh, MM - Middle Mtkvari. Mineral deposits: 65 - Apshra, 66 - Amakekva, 67 - Adzaga, 68 - Chordi, 69 - Pitikvara, 70 - Kudaro, 71 - Khaishi, 72 - Gumbri, 73 - Askana, 74 - Chorchana group, 75 - Dzegvi, 76 - Tedzami, 77 - Kisatibi, 78 - Paravani, 79 - Tsikhisjvari, 80 - Gldani, 81 - Grdzeli Tba, 82 - Sakhare Tba, 83 - Kutaisi, 84 - Lechkhumi, 85 - Algeti, 86 - Taribani, 87 - Adjameti, 88 - Brdzyshkha, 89 - Chititskaro, 90 - Matkhodji-Udzlouri, 91 - Shrosha-Ubisi, 92 - Agara, 93 - Shrosha, 94 - Jvarisi, 95 - Akhaltsikhe, 96 - Dmanisi, 97 - Goderdzi, 98 - Kayundag.

Figure 7. Distribution of Deposits of Chemical, Ceramic Materials and Gems in Georgia

Appendix.

Iron deposits are related to 4 mineralization epochs. Early Alpine hydrothermal activity of the Bolnisi Mining District (Somkhito-Karabakh terrane) just started by formation of iron ores. Middle Alpine sedimentary iron occurrences are due to coal-bearing strata of the Gagra-Java subterrane. Late Alpine skarn-magnetite mineralization is connected with granodiorite-monzonite complexes in the western part of the Adjara-Trialeti subterrane. At least, littoral zone of the Dzirula subterrane hosts modern iron-bearing sea placers.

The Dzama skarn-magnetite mine in the Adjara-Trialeti subterrane is related to an exocontact of the Neogene gabbro-diorite Kvirani massif. Magnetite skarns contain massive (Fe content 45-60%), high-grade (Fe content 30-45%) and low-grade (Fe content 20-30%) impregnation ores [20].

The Poladauri hematite bodies in the Bolnisi Mining District were mined from 1862 up to 1927. Hematite ores are related to Upper Cretaceous volcanic and tuffaceous rocks and, according to the recent ideas on ore-zoning in this district [12], belong to the lower part of the ore-metasomatic column in ascending section being rapidly replaced by copper and barite-polymetallic ores. The mine consists of several strata-bound ore bodies with average thickness of 12 m. Fe content varies from 32 up to 41% but ores are easily enrichable.

The Tkibuli-Shaori deposit is related to the Middle Jurassic Tkibuli-Shaori coal basin mentioned above. Siderite strata alternate with coal seams here. Mean grade of siderite ores is 34.5%, average thickness of ore bodies – 3.9 m.

Magnetite placers of the Black Sea coast are 50 km long, but they are extremely rich near the influx of the river Supsa (Supsa-Natanebi deposit). Here content of magnetite and titanomagnetite fractions is 3% with reserves of 45,000,000 tons (15,400,000 tons of iron).

From the XIX century Georgia has been known as having important reserves of manganese oxide and carbonate ores. All commercial deposits and mines are due to the Middle Alpine mineralization epoch. In Oligocene time the Maikop basin was subject to typomorphic dyagenetic and autigene mineral formation [28, 29, 37] comprising glauconite, chalcedony, manganese minerals. By that time large carbonate and mixed carbonate-oxide ores were deposited at the northern and south-western edges of the Dzirula massif (Chiatura, Kvirila, Chkhari-Adjameti basins, etc.). These ores underwent entire or partial dyagenetic and autigene alteration up to oxide and peroxide ores. Calculations of mineral balance [14] indicated that Mn washed out from the Greater Caucasus and the Dzirula weathered granites was insufficient for the mineralization, and postvolcanic hydrothermal supply of this metal is to be taken into account [28, 29]. Participation of the hydrothermal matter in the sedimentation, dya- and epigenesis processes is outlined by typomorphic mineral composition of Mn-bearing layers: opal, glauconite, zeolites (clinoptillolite), montmorillonite with only minor terrigenous fraction [33].

Table IV illustrates official data on ore grade and reserves in main manganese deposits of Georgia. Thus, Mn reserves are limited, and the Chiatura Mine can maintain its status of the main metal supplier for the ferromanganese plant in Zestaphoni only for 10-20 years.

TABLE IV: Manganese Reserves in Main Georgian Deposits [50].

Deposit	Ore Type	Mn grade	Reserves, thousand t			
			1969-72		1995	
			1969-72	1995	1969-72	1995
Chiatura	Oxide	27.8	96,970	26,958.7	84,376	23,456.5
	Peroxide	33.4	44,289	14,792.5	24,377	8,141.9
	Carbonate	16.6	90,909	15,090.9	90,909	15,090.9
Chkhari-Adjameti	Oxide	24.8	3,756	931.5	3,756	931.5
	Peroxide	28.2	3,574	1,007.9	3,574	1,007.9
	Carbonate	15.8	2,709	428.0	2,709	428.0
Kvirila	Oxide	23.3	3,000	699.0	3,000	699.0
	Carbonate	19.2	24,000	4,608	24,000	4,608.0
Shkmeri	Oxide	28.9	1,225	354.0	1,225	354.0
	Carbonate	20.7	5,173	1,070.8	5,173	1,070.8

5.4. PRECIOUS, NON-FERROUS, LIGHT, RARE METALS AND ELEMENTS*

Excluding small placers in Enguri and Khrami river basins (not more than 8,000 kg Au in both of them), Au is mainly related to:

1. Base metal massive sulphides;
2. Porphyry copper systems.

Two types of base metal massive sulphides are widespread in Georgia [49]. The first one is related to typomorphic Cimmerian (Lower Jurassic) mineral formation in the Chkhalta-Tphani subterrane and consists of copper-pyrrhotite stratiform and vein ores (Adange, Zeskho, Artana). These thoroughly investigated deposits [11, 43, 44, 51] represent hydrothermal-sedimentary copper massive sulphides, which underwent replacement of pyrite by pyrrhotite during regional green schist metamorphism. According to existing data [11, 43, 44] these copper-pyrrhotite deposits, in common, consist of three zones: (i) main ore zone presented by a concordant pyritic stratum metamorphosed into copper-bearing pyrrhotite; (ii) subore zone consisting of copper-pyrrhotite veins suggested to be metamorphic ore channels; (iii) supraore zone built up by non-commercial quartz-pyrite veinlets interpreted as a metamorphic halo. This halo is suggested to represent a result of sulphur removal from stratiform ore bodies during "pyrrhotinization" of pyrite. It is interesting to note that there is a direct relation between composition of pyrrhotite (including its sulphur isotopic composition) and type of ores (inherited stratiform or vein). If the former bodies are represented by typical hexagonal pyrrhotite $Fe_{10}S_{11}$, these veins consist of admixture of pseudohexagonal and monoclinic pyrrhotites that have a varying composition from Fe_6S_7 up to Fe_8S_9 [51]. This event may be explained by kinetics and thermodynamics of interaction of pyritic bodies and cupriferous hydrothermal-metamorphic solutions [43]. Copper supply during metamorphism

---

* See in Figure 6.

of pyrite bodies is proved by: (i) constant difference in *Cu* content in vein and stratiform ores; (ii) absence of *Cu* in supraore zone; (iii) constant difference in *Cu* content in relict pyrite and inherited-stratiform *Cu*-pyrrhotite ores; (iv) crystallographic and experimental data demonstrating obligatory metamorphic "shrinkage" in the course of "pyrrhotinization" process [44]. The latter process is due to the fact that the volume of elemental cell of pyrite (about 0.1584 nm^3) is more than that of hexagonal pyrrhotite – 0.05829 nm^3 [44]. Experimental data [52] demonstrated that pyrrhotite formed as a result of thermal decomposition of pyrite occupies by 30% less volume than the iron disulphide. As inherited-stratiform *Cu*-pyrrhotite ores have a massive structure, it seems logical to suppose that free space was occupied by quartz-non-ferrous paragenesis. Calculation of matter balance and comparison of these data with *Cu* content in different ores indicate that about 25% of *Cu* was supplied during metamorphism.

Three large deposits of this type have only a minor gold content but are characterized by important copper reserves (see Appendix).

The second group of base metal massive sulphide deposits is related to the Bolnisi Mining District. These deposits were formed during the Early Alpine mineralization epoch within the limits of the Somkhito-Karabakh terrane. *Au-Ag-Cu-Pb-Zn-barite* deposits are connected with calc-alkaline volcanic depressions and represent metasomatic analogues of the Kuroko-type [21] ores [27]. Deposits are characterized by vertical zoning [22, 37]: in the ascending section *Cu* ores are replaced by lead-zinc and then barite mineralizations. Commercial contents of *Au* are related to both *Cu* (pyrite-chalcopyrite) ores and ore-hosted quartz-alunite-sericite metasomatic rocks. Official data on ore and metal reserves in the Madneuli Mine (the largest deposit of this type) are given in Table V. Metal reserves of other important deposits (Tsiteli Sopeli, David-Gareji, Kvemo Bolnisi, Sakdrisi) of the Bolnisi Mining District are shown in Appendix.

TABLE V. Economic Characteristics of the Madneuli Mine (after [50])

Ore Type	**Element or Mineral**	**Grade, %**	**Reserves, t (kg for Au)**		
			Ores	**Element or Mineral**	**Already Mined Element or Mineral**
Cu	Cu Au Ag S Se Te	1.28 0.000073 0.000431 6.92 0.000710 0.000759	79,087,500	516,700 5,773 34 5,472,855 56,152 60,027	67,700 756 5 717,074 7,357 7,865
Cu-Zn	Cu Zn	0.39 1.80	1,425,200	5,558 25,654	
barite-polymetallic	barite Pb Zn	31.7 0.62 4.09	1,324,000	419,708 8,209 54,152	125,306 5,900 10,600
barite	barite	36.3	338,000	122,694	122,694
Au-bearing quartzites	Au Ag	0.000180 0.000915	10,896,000	48,247 100	

Only the Madneuli Mine is exploited both by the Joint Stock Company "Madneuli Metal Enrichment Plant" (producing gold-bearing copper concentrate) and the Georgian-Australian joint venture "Quartzite, Ltd." (producing gold from secondary quartzites). In close vicinity to these deposits the Dambludi veins host 1,882 kg of *A*u with grade of 1.18 ppm; 50.63 tons of *Ag* (30.10 ppm); 13,700 tons of *C*u (0.73%); 47,600 tons of *P*b (2.5%); 95,900 tons of *Zn* (5.3%); 49.5 tons of *In* (26.5 ppm).
The Merisi Mining District is much smaller. It is related to the Adjara-Trialeti subterrane, which originated during Palaeogene as a rift zone and is composed by a volcanic-plutonic affinity. Merisi-Namonastrevi granodiorite-monzonite complex in Mountainous Adjara is surrounded by a number of polymetallic quartz veins, which form a peripheric zone of a porphyry copper system [53]. Polymetallic lodes have been mined since the beginning of our century. Mining was suspended during the Second World War. Present-day reserves of this district are shown in Appendix.
In addition to these typical auriferous ore formations, about 10 tons of gold (with grade of 0.5-3.5 ppm) and at least two times more of economic resources are known in the Lukhumi *Au-As* Mine, characterized by very unusual geological features [54]. This Mine belongs to the Carlin-type deposits, where realgar-orepigment mineralization is encountered in a thick stratum of limestones among rhythmical calcareous Upper Cretaceous affinity. The ore zone is related to a long-living deep fault at the margin of the Mestia-Tianeti and the Chkhalta-Tphani subterranes. Gold-bearing arsenopyritic impregnation forms a halo between the handing wall of the fault and the lying contact of realgar-orepigment veinlet bodies. Grade of ores and metal reserves are shown in Appendix.
Beyond the Madneuli and the Merisi mining districts, there are 8 lead-zinc deposits in Georgia. All of them are distributed within the limits of Gagra-Java subterrane and were formed during the Early Alpine mineralization epoch. These deposits belong to two genetic groups: (i) Late Jurassic – Early Cretaceous (?) Mississippi Valley-type limestone-related ores [20], which overlap the roof of Jurassic evaporates in the North-Western (Brdzyshra and Dzyshra); (ii) Cretaceous telethermal sphalerite-galena-quartz-calcite subvertical lodes of enormous and large vertical extension (over 1,500 m at the Kvaisi Mine, about 300-500 m at other deposits: Skatykom, Razdarankom, Amtkheli, etc.). These veins have important admixture of *Ag, Ga, Ge, Se, Te, Tl* [61]. Among them the largest is the Kvaisi Mine, whose reserves are given in Table VI. Resources of important *Ga, Ge, Tl, Se, Te* admixtures have not been calculated[4]. Reserves of other deposits are shown in Appendix. Total reserves of precious and non-ferrous metals in Georgia are shown in Table VII.

---

[4] These elements were reduced to the Pb-Zn concentrate and transported to the "Elektrozink" Plant, Vladikavkaz (former Ordjonikidze), Ossetia, Russian Federation, where their ingots were produced without any payments to the mine.

TABLE VI: Economic Characteristics of the Kvaisi Mine [50]

Metal	Grade, %	Ores	Reserves, t	
			Metal	Metal Already Mined
Pb	2.24	2,851,000	56,200	37,600
Zn	6.47		197,700	134,200
Ag	0.001540		18	34
Cd	0.02		279	418

TABLE VII. Bulk Reserves of Precious and Non-Ferrous Metals in Georgia

Metal	Bulk Reserves, t
Au	98
Ag	784
Cu	2,403,000
Pb	324,560
Zn	816,821

There is a number of rare metal/element deposits in Georgia, with few of them having large tonnage and high grade of ores. These deposits are distributed in different terranes. Some features of the Lukhumi Mine were mentioned above. Specific mineralization processes related to the formation of the Greater Caucasus mountainous system determined origination of other deposits, distributed in the Greater Caucasus terrane and the Chchalta-Tphani subterrane. These deposits occur either near ore-bearing orogenic granitic domes (*Mo, W*) or in fault-related fissure zones (*Hg, Sb,* partly *As*). Reserves and grades of these deposits are shown in Appendix.

An unusual Gelati *Al* deposit must be mentioned separately. *Al* presence is determined by specific Early Alpine autigene ore formation in the Dzirula subterrane. The deposit represents a 31.2-114.5 m-thick stratum of analcite sandstones composed of *Al*-bearing analcite, quartz, feldspar, mica, pyrite and calcite. Analcite occupies about 80-90% of the sandstone matrix and content of Al_2O_3 reaches 18.9%. Reserves of these ores in the district, explored only partially, are over 300,000,000 tons, whereas the Gelati deposit itself hosts 176,000,000 t of ores.

5.5. CHEMICAL & AGROCHEMICAL RESOURCES*

Georgian chemical and agrochemical resources imply barite, bentonites, talc, zeolites, diatomite, perlite, acid-resisting andesites, phosphorites, Glauber salt, lithographic stones, haloids, chalcedony and mineral paints.

---

* See in Figure 7.

5.5.1. *Barite*

There are three types of barite deposits in Georgia: (i) barites as the upper oxidizing zone of the base metal massive sulphides in the Bolnisi Mining District [27]; (ii) Early Alpine thick barite lodes in the Gagra-Java subterrane [45]; (iii) also Early Alpine strata-bound barite deposits in Upper Jurassic limestones, overlapping an evaporite sequence [8] in the same subterrane. The bulk reserves of five most important deposits (Madneuli, David-Gareji and Kvemo Bolnisi in the Bolnisi Mining District, Chordi vein and Apshra strata-bound deposits) contain 7,722,000 tons of barite for chemical use and paints including 5,820,000 tons of flotation barite (3,200,000 tons in the Bolnisi mining District and 2,650,000 tons in the Apshra deposit). Grade and reserves of all deposits are shown in Appendix.

5.5.2. *Bentonite*

Georgian montmorillonitic clays were introduced into chemical and petrochemical industries by my grandfather, Academician Alexander A. Tvalchrelidze. He started his investigations in 1916 and just that summer registered 53 manifestations of these clays, analogous to the fuller earth from Georgia and Florida States, USA. In early 1927 he organized a comparative investigation of kerosene purification by montmorillonitic clay from Gumbri, Georgia, and by fuller earth from Attapulgus, Georgia, USA at the US concession plant in Baku, Azerbaijan, by that time owned by Standard Oil Company of New York. These industrial tests have proven high quality and excellent technologic properties of the Georgian montmorillonitic clays, which received the brand mark of "gumbrine" [41]. Geologically gumbrine represent a product of autigene marine (halmirolysis) alteration of the andesitic ash. This conception, first forwarded by Tvalchrelidze (1941), was afterward proven by Machabeli [32, 33]. Among numerous manifestations of gumbrine the better technologic properties are characteristic of the Gumbri and the Askana group deposits (Early Alpine mineralization of the Dzirula subterrane). Reserves are equal to about 11 millions tons of this precious raw material.

5.5.3. *Talc*

Bulk current talc reserves of the Chorchana group deposits amount to 1,900,000 tons. These deposits are due to the Chorchana ophiolitic hyperbasite Hercinian complex of the Dzirula crystalline massif.

5.5.4. *Zeolites*

Georgian zeolites played important role in stopping nuclear chain reaction during the Chernobyl catastrophe. In normal circumstances they are used in agriculture, as oil filters and, seldom, as molecular cells. Two especially rich (more than 60% of clinoptillolite) deposits (Dzegvi and Tedzami) have bulk reserves of 37 millions tons [50]. Zeolite mines are distributed at the eastern edge of the Late Alpine Adjara-Trialeti subterrane and are determined by specific post-volcanic lacustrine sedimentation and autigene mineral formation with a possible hydrothermal supply of silica [22].

5.5.5. *Diatomite*

Diatomite of the Kisatibi Mine, near the town of Akhaltsikhe, was formed in shallow fresh-water basin, which originated in Late Alpine time within the limits of the Adjara-Trialeti subterrane. According to the existing model, SiO_2 was supplied by postvolcanic hydrotherms of the Adjara-Trialeti Late Alpine volcanism. Physical and chemical properties of the Kisatibi diatomite [20] are among the best in the world (SiO_2 91-96%, Fe_2O_3 0.28-1.5%, density of dry fraction 0.56-0.83 g/cm^2, amount of diatom tests over 20 millions per cm^3). Actual reserves are over 7,800,000 tons. Diatomite is mined since 1950 for numerous industrial branches including production of thin oil filters.

5.5.6. *Perlite*

Perlite of the Paravani deposit is due to Neogene volcanic processes of the Somkhito-Karabakh terrane. Its reserves are over 60 million m^3. Perlite is used in numerous branches of industry, including production of insulators and of perlitoconcrete. Technologic properties are briefly described in Appendix.

5.5.7. *Acid-Resisting Andesites*

The Tsikhisjvari deposit of acid-resisting andesites represent a 35 m thick aerial flow of the Early Quaternary Sairmos Kedi volcano. Thin grinded andesites have excellent acid-resisting properties. They are used in different branches of industry comprising chemical and petrochemical processes and production of acid-resisting equipment [20]. Reserves of andesites are evaluated to be 5,194,000 m^3.

5.5.8. *Phosphorites*

Phosphorite occurrences in Georgia are related to the Middle Alpine mineragenesis in the Gagra-Java subterrane. In Oligocene time several shallow water basins were subject to intense deposition of fish skeletons due to volcanic paroxysms. Among phosphorite occurrences only the Kutaisi and the Lechkhumi deposits have moderate commercial interest. Lechkhumi Oligocene phosphorite deposit is encountered at the bottom of a glauconite sandstone sequence. Reserves of phosphorites with P_2O_5 grade over 10.8% amount to 407 thousand tons. Main economic indices of phosphorite deposits are stipulated in Appendix.

5.5.9. *Glauber Salt*

Two types of Glauber salt (mirabilite) deposits are known in Georgia within the limits of the Middle Mtkvari subterrane [48]: (i) modern salt lake sediments (Grdzeli Tba, 608,000 tons, Sakhare Tba, 386,000 tons) and (ii) Quaternary evaporites (Gldani deposit, rest reserves of 65,000 tons).

5.5.10. *Lithographic Stones*

The unique Algeti sedimentary Late Cretaceous deposit of a lithographic stone, used as inert container matrix in artificial diamond-producing industry, has reserves of 1,625,000 m^3. Its technologic properties are given in Appendix.

5.5.11. *Haloids*

The Taribani deposit of haloids represents oil-related waters of an oil field, which at a depth of 2-2.5 km are rich in these elements. Their average contents are equal to (in mg/l): I - 20.3; Br - 14.0; B_2CO_3 - 23.5.

5.5.12. *Chalcedony*

The Adjameti chalcedony deposit was formed by dyagenetic-autigene alteration of the Oligocene spongolite strata that replace westward from the Chiatura town the manganese-bearing sequence. It is thought [29] that the spongolite strata fix margins of manganese-bearing basins. Chalcedony contains 96.20 - 98.56% of SiO_2. Heat resistance is 1730-1750°C, density - 2.60-2.62 g/cm^3. Reserves of chalcedony are 4,090,000 tons and those of spongolites – 16,182,000 m^3.

5.5.13. *Mineral Paints*

Mineral paint deposits of Georgia mainly include different ochres (red, pink, yellow, blue, etc.) and limonite pigments. The deposits are related to weathering processes of either polymetallic deposits (Brdzyshra) or granitic and other rocks rich in $Fe_2O_3 + FeO$ (Matkhoji-Udzlouri and Agara). Reserves of these deposits and their technologic properties are given in Appendix.

5.6. CERAMIC RESOURCES*

Resources of ceramic materials are fairly limited in Georgia. They are related to: (i) Palaeozoic pegmatites of the Dzirula crystalline massif; (ii) Early alpine kaolinite deposition in the Dzirula subterrane. Current reserves of the Shrosha pottery pegmatites are 2,200,000 tons. Porcelain Middle Jurassic white kaolin clays at the Jvarisi deposit do not exceed 1,300,000 tons. Recently [35], however, usage of the inner zones of quartz-alunite-sericite hydrothermally altered rocks at polymetallic massive sulphide deposits (Tsiteli Sopeli, for instance) instead of China stone has been proven to be possible.

5.7. GEMS*

Georgia is not characterized by a great variety of jewellery gem deposits but the existing manifestations of agate, opal and obsidian are distinguished by picturesque texture and very good jewellery properties. Two main deposits of agate (Akhaltsikhe in the Adjara-Trialeti subterrane and Dmanisi in the Somkhito-Karabakh terrane) are related to fault structures in Eocene volcanic sequences filled by opal-chalcedony hydrothermal matrix. Agate form numerous nests (their mass is often several kg) of red, rose, pink, blue colours and the thin-layered structure. Bulk reserves are over 3 million kg [48]. Obsidian

* See in Figure 7.

from the Kayundag deposit (a Neogene rhyolite flow) has excellent technical characteristics for different kinds of art and jewellery productions.
A unique geological phenomenon - the Goderdzi deposit of opalized wood must be mentioned (Adjara-Trialeti subterrane) separately. Accumulation of wood was determined by shallow-water bog sedimentation buried under Late Neogene andesitic pyroclastics. Mineralization represents practically a pseudomorphosis of opal upon the wood branches and trunks. Opalization occurred due to rapid dyagenetic and autigene alteration under influence of postvolcanic hydrothermal SiO_2 supply. The 600-m-thick opal-bearing sequence extends for 15 km, covering an area of 85 km^2. Entirely opalized wood trunks often form large accumulations. The texture of yellow to dark brown and even black opal repeats in every detail primary composition of the replaced wood. Opal has magnificent decorative properties but the deposit must be exploited very carefully in the course of protection of this natural phenomenon.

5.8. FACING MATERIALS*

As far back as the Middles Ages Georgia, like Greece and Italy, was known for its vast supplies of facing materials. The Kremlin Convention Hall, International Trade Center, many other buildings and subway stations in Moscow, St. Petersburg and numerous other towns of Russia are adorned by Georgian granites, teschenites, marbles and tuffs. Georgian facing materials have excellent decorative features and corresponding physical and mechanic properties (Table VIII). Their reserves are very important. Appendix contains information on technologic properties and reserves only of main deposits which are presented by [50]:

- Typomorphic pre-Cambrian gabbros of the Dzirula massif with pink plagioclase monocrysts known as «rikotite» (Rikoti deposit, 4,918,000 m^3);
- Palaeozoic granites and quartz diorites of the same massif (Rkvia and Tsipi deposits, over 50 millions m^3);
- Middle Jurassic teschenites (specific analcite-bearing diabase) of the Dzirula subterrane (Kursebi and Opurchkheti deposits, over 7,500,000 m^3);
- Numerous dikes of gabbro-diabases in different zones with reserves beyond limits;
- Basalts from dolerite ingressive flows in Southern Georgia (Tsalka, Ratevani, Marneuli, etc. deposits with reserves beyond limits);
- Upper Cretaceous decorative tuffs of the Somkhito-Karabakh zone (Bolnisi group, and other deposits, about 27 million m^3);
- Marbles in the Triassic Dizi series of the Chkhalta-Tphani subterrane (Dizi, Chuberi, Choluri, 4,750 thousand m^3);
- Over 20 deposits of marmorized limestones of different age (Salieti, Moliti, Lopota group, etc., over 42 millions m^3).

---

* See in Figure 8.

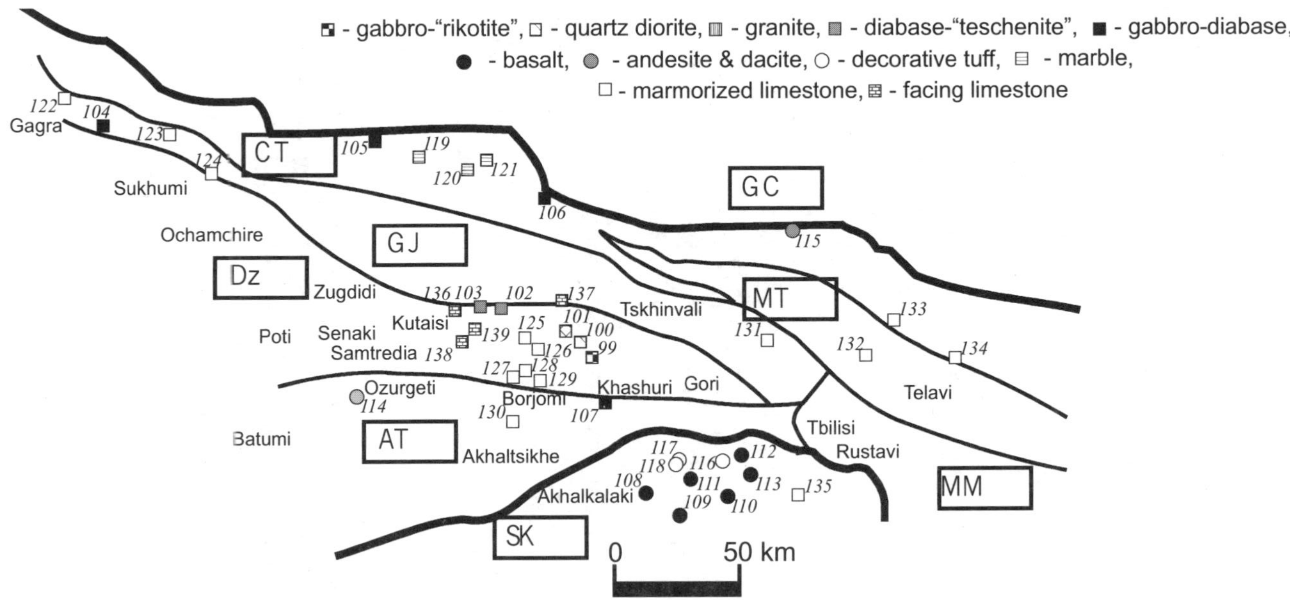

Terranes and subterranes: GC - of the Greater Caucasus, CT - Chkhalta-Tphani, MT - Mestia-Tianeti, GJ - Gagra-Java, Dz - Dzirula, AT - Adjara-Trialeti, SK - Somkhito-Kakabakh, MM - Middle Mtkvari. Mineral deposits: 99 - Rikoti, 100 - Tsipi, 101 - Rkvia, 102 - Kursebi, 103 - Opurchkheti, 104 - Aigbi, 105 - Chirkvali, 106 - Kvaishauri, 107 - Ushlevi, 108 - Tsalka, 109 - Beshtasheni, 110 - Kazreti, 111 - Ratevani, 112 - Marneuli, 113 - Karadag, 114 - Shemokmedi, 115 - Kobisi, 116 - Bolnisi group, 117 - Chivchavi-I, 118 - Chivchavi-II, 119 - Dizi, 120 - Chuberi, 121 - Choluri, 122 - Kldiana, 123 - Gumista, 124 - Shroma, 125 - Salieti, 126 - Moliti, 127 - Dzveli Shrosha, 128 - Akhali Shrosha, 129 - Merelisi, 130 - Chobareti, 131 - Kanchaveti, 132 - Jakhuri, 133 - Lopota, 134 - Ilto, 135 - Sadakhlo, 136 - Eklari, 137 - Darkveti, 138 - Kvakhchiri, 139 - Motsameta.

Figure 8. Distribution of Deposits of Facing Materials in Georgia

TABLE VIII. Main Physical & Mechanic Features of Some Georgian Deposits of Facing Materials

Deposit	Density, g/cm³	Porosity, %	Water absorption, %	Ultimate resistance in dry state, kg/cm²	Softening coefficient	Frost resistance coefficient	Attrition, g/cm²
Rikoti, rikotite	3.0-3.2	<1	0.2-0.5	1176-1475	0.86-097	0.88-1.00	0.20-0.30
Kursebi, teschenite	2.7-2.9	2-6.8	0.2-1.7	1257-1945	0.71-0.93	0.89-0.90	0.31-0.40
Marneuli, basalt	2.8-3.0	10-27	0.8-1.3	402-884	0.83-0.96	0.75-0.99	0.58-0.71
Bolnisi group, tuffs	2.3-2.7	10-12	0.8-8.9	867-1100	0.85	0.90	1.33
Dizi, marble	2.7-2.8	1.5	0.6-0.9	546-850	0.80-0.87	0.31-0.49	1.45
Lopota group, marble	2.7-2.8	2.0-2.4	0.2-0.3	663-773	0.72-0.84	0.97	3.60-4.00
Salieti, marmorized limestone	2.7	2.9	0.5	1020	0.89	0.90	2.50
Moliti, limestone	2.7-2.8	0.8-2.9	0.1-0.4	530-849	0.95	0.95	3.80

5.9. INERT, METALLURGIC & BUILDING RESOURCES*

5.9.1. *Inert Materials*

Among inert materials flux limestones, sand & gravel, sands for glass industry and foundry sands may be distinguished [34]. Appendix briefly describes their technologic properties and reserves.

Flux limestones must have chemical ($CaO > 50\%$; $SiO_2 < 3.9\%$) and physical (resistance to compression 1,200-1,600 kg/cm²) properties available for using in metallurgy (Rustavi and Zestaphoni plants in Georgia). These numerous occurrences are mostly related to Cretaceous carbonate sequences of the Dzirula and the Middle Mtkvari subterranes. Bulk reserves only of the most representative deposits are over 75,800,000 tons [20].

Exploitation of the Black Sea beach gravel deposits is prohibited because of an ecological risk. However, numerous pre-Quaternary and Quaternary gravel and sand accumulations are known in the Mtkvari, the Alazani and other river valleys their bulk reserves exceeding 1 billion m³.

Deposits of foundry sands differ from those of sand for glass industry by elevated feldspar content and, correspondingly, low quartz percentage. The most widespread occurrences of either type are related to the Oligocene sequence both near the Chiatura manganese deposit and not far from Tbilisi outcrops of the Dzirula crystalline massif. These sands are presented by two types: (i) pure quartz sands with SiO_2 content of 93-94%; and quartz-feldspar sands with SiO_2 content below 90%. Yield of the fraction enriched in SiO_2 (up to 94%) is about 60% [20].

* See in Figure 9.

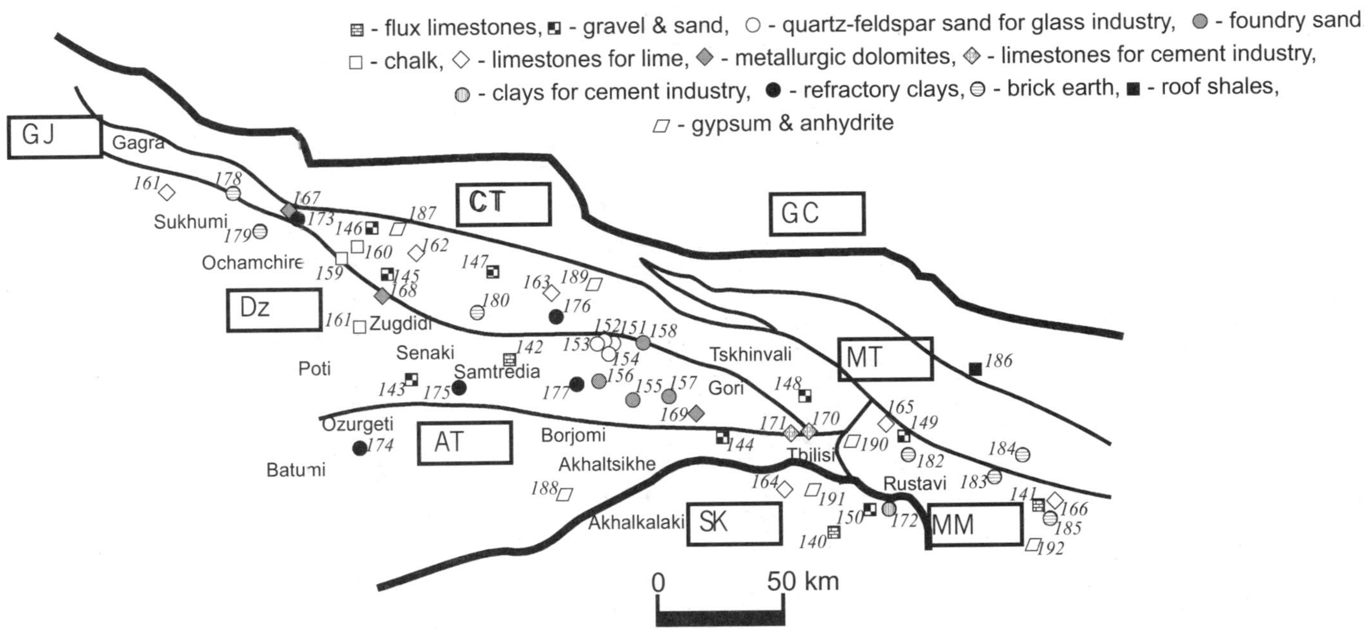

Figure 9. Distribution of Deposits of Inert, Metallurgic and Building Materials in Georgia

5.9.2. *Metallurgic Materials*

In addition to flux limestones metallurgic industry needs certain types of dolomites having enhanced refractory capacity (> 1600°C). Metallurgic dolomites are due to the Lower Cretaceous calcareous sequence of the Dzirula and the Gagra-Java subterranes. Their bulk reserves are over 60 million m^3.

5.9.3. *Building Materials*

Deposits of this type comprise a lot of industrial minerals including chalk, limestones for lime production, limestones and clays for cement production, refractory clays, brick earth, roof shales, gypsum, anhydrite, etc.

Chalk deposits are related to the Gagra-Java subterrane and have the Late Cretaceous age. All of them have $CaCO_3$ content above 95%. Appendix contains data only on the reserves of two most important mines.

Limestones for lime production must have active $CaCO_3 + MgCO_3$ above 75%, they must be slaked under temperatures of 1000-1200°C with rate under 30 min, have a yield above 1 l/kg and amount of unslaked particles below 30%. Such deposits are distributed in different subterranes, predominantly have either Cretaceous or Neogene age and are characterized by considerable reserves. Appendix contains information only on several mines.

Among numerous occurrences of limestones for cement industry only tow deposits with bulk reserves over 110 million t are shown in Appendix. These deposits are absolutely sufficient for 50 years under conditions of cement production rate being 5-10 million t per annum. Deposits are situated in economically developed area and have shown excellent technologic properties. All deposits of this type belong to Cretaceous and Neogene limestone sequences. Among the numerous alluvial deposits of clays for cement industry the Gardabani Mine is distinguished by excellent properties: SiO_2/R_2O_3+RO – 2.26; Al_2O_3/R_2O_3+RO – 2.60, density – 1.60-1.90 g/cm^3.

Very important deposits of refractory clays are due to either Neogene red-clay formation of the Adjara-Trialeti and the Dzirula subterranes (tuffaceous, slightly weathered rocks) or Lower Jurassic schists at the periphery of the Dzirula massif. They differ from country rocks by high melting temperature (above 1700°C).

Among approximately 600 manifestations of brick earth 106 have important dimensions and reserves. Thus, we have shown only the largest deposits in the register and on the map (see Figure 10). All deposits are presented by alluvial lowland clayey sediments of the largest Georgian rivers. Deposits are characterized by refractory capacity above 1000°C, foundry moisture of 10-45%, shrinkage of 9-30%, and water adsorption above 3%. Bulk reserves of 106 deposits are 76,161,000 m^3 [20].

The Kakheti (Intsoba) group of roof shale deposits is composed by Lower Jurassic black schists of the Chchalta-Tphani subterrane. Three strata of roof shales have a bulk thickness of 138 m. Yield of tiles (0.2-1.8 m^2 with thickness of 6-80 mm) is 4.8% [20].

The great majority of gypsum and anhydrite deposits are due to lacustrine evaporitic sediments (Upper Jurassic, Neogene) or to their weathering core. In addition, there is a vast number of alluvial, diluvial and proluvial modern deposits containing relatively low percentage of anhydrite. These deposits are used for preparation of industrial solutions utilized in interior facing works. Reserves and grades of the largest deposits are shown in Appendix.

5.10. MINERAL WATERS

Georgia is well known for its sources, outsets and artesian basins of mineral waters characterized by unique properties and enormous reserves. Against the background of both very limited perspectives of developing water resources in Europe and increasing necessity in underground waters, Georgia is able to provide outside markets with any needed quantity of high quality waters. The main water reserves are distributed in picturesque, ecologically unspoiled valleys, thus providing one of the most important prerequisites for future development of spas.

Georgian underground waters are usually classified into four groups [48, 49]:

- Fresh waters (mineralization degree < 1000 mg/l) have reserves of 550 million m^3 per day (23 million m^3 per hour). Reserves are due to vast artesian basins of both the Western and the Eastern Georgia (the Kolkheti and the Eastern Georgia Lowlands). Their chemical, physical and bacteriological tests show them to be superior to the world-wide best quality waters. At the same time, their utilization (including tap waters) is about 1-2%.
- Hydrocarbonate, mainly sodium and calcium table waters (mineralization degree from 1-5 g/l) have bulk reserves of 3,000 m^3 per day (125 m^3 per hour). They are due to either postvolcanic infiltration processes of the Adjara-Trialeti mountains or to fault-related activity along relatively young tectonic dislocations. World demand for such waters is very high whereas supply is limited.
- Table-medicinal waters of the same origin and chemical composition but with important admixture of gaseous phase (CO_2, CH_4OH) have mineralization degree 5-11 g/l. They are exploited for both external and internal usage. The well-known Borjomi spa contains just these waters. Their bulk reserves are 13,000 m^3 per day (542 m^3 per hour).
- Medicinal sulphide, silica, *Ra, N* waters of different cation composition and mineralization degree are used for medicinal purposes only. Their Bulk reserves are 30,500 m^3 per day (1,271 m^3 per hour). Well-known medical spas like Tskhaltubo and Tbilisi just belong to this type.

6. Economic Evaluation of the Mineral Resource Base

Table IX contains information on main types of Georgian mineral resources, registered by corresponding state services and re-calculated according to the above-mentioned approach. It must be noted that a degree of industrial utilization is calculated as a sum of ratios of the rest and primary reserves price indices of all deposits.

TABLE IX. Macroeconomic Evaluation of the Georgian Mineral Resource Base

N°	Main Types of Mineral Resources	Comparative Economic Importance, %	Degree of Industrial Utilization, %	Price Index of Rest Reserves, $ mln
1	Fuel & Energy Resources	33.6	40	30 240
2	Ferrous Metals	10.1	45	9 090
3	Non-Ferrous & Precious Metals	12.4	6	11 160
4	Rare Metals, Elements & Earths	0.8	5	720
5	Chemical & Agrochemical Materials	4.7	15	4 230
6	Ceramic Materials	0.5	70	450
7	Gems	0.9	3	810
8	Facing & Decorative Materials	8.6	10	7 740
9	Building Materials	4.0	25	3 600
10	Underground Mineral & Thermal Waters	22.0	4	19 800
11	Other Resources	2.4	35	2 160
12	TOTAL	100.0	24	90 000

Thus, Georgia processes almost all needed mineral resources, which often have unique technologic properties. In case of efficient development of mining and processing businesses Georgia will be able both to satisfy demands of industry for raw materials and products of their processing and to organize export of industrial minerals and precious and non-ferrous metals. However, the attainment of this objective needs entire reform of the governmental managerial chart, when instead of development of concrete kinds of resources governmental support would be given to entire capitalization and marketing of the resource base as such. And, first of all, the State reserve of the resource base should be created. Hence, such a goal must be based on numeric macroeconomic model of the Georgian mineral resource base.

7. Perspectives of the Georgian Mineral Resource Base

In 70s and 80s of the XX century a specific paradoxical model of economic investigations was widespread in the whole USSR: prognosis investigations on various mineral resources covered the entire territory of USSR. The notion of *"prognostic reserves"* was introduced implying a possible amount of an unexplored mineral resource on a given territory. Moreover, a corresponding mathematical apparatus was elaborated, and such an apparatus tried to describe in formal terms informal information on distribution regularities of mineral resources in the earth's crust [36] (quantitative metallogeny, etc.). However, from the very beginning, such approach had a fundamental fault: prospecting and exploration strategy was never based on their *economic efficiency* or, in other words, necessity to discover such reserves of mineral resources, which would ultimately be recoverable. The soviet economic school took into account the *geologic efficiency* of prognosis, which implied absolute amount of mineral resources that ultimately could be discovered by exploration. Of course, this feature was entirely determined by the soviet determination of the resource base, mentioned above. It is not difficult to conclude that almost all this information is useless for a country, which tries to introduce free market economic relations. That is why in early 1990s I tried to re-interpret the existing information on Georgian resource base and to create the macroeconomic model of its possible development [49].

Mineral resources appraisal by models of the economic activity implies the relationship between the volume of exploration and the quantity of mineral resources, which ultimately will be recoverable [12]. Within the framework of the economic terms this quantity is equivalent to potential supply. Pioneer of such investigation was Hubbert [25] who performed *a premier case study* to determine the quantity of crude oil, which ultimately would be recovered in the United States of America, and, thereby, elaborated a macroeconomic method that lead to political solution of the energy crisis.

According to Hubbert, in terms of the *"discovery-rate"* model the quantity of a recoverable mineral resource is a function of the cumulative footage of the exploration drilling:

$$Q = f(h), \tag{4}$$

where Q = reserves of a mineral resource and h = footage of the exploration drilling. Then, $f(\infty)$ would correspond to the ultimate recoverable reserves Q_∞. Since the earth is finite and mineral resources in the deposits provide sizeable target, Q_∞ can be approximated by evaluating f as some finite footage $\tilde{h} < \infty$:

$$\begin{array}{l} f(\tilde{h}) \to \infty \\ Q \to Q_\infty \end{array}. \tag{5}$$

Statistical evaluation of these regularities is possible if we consider another function $\varphi(h)$ as derivative of the function $f(h)$:

$$\frac{df(h)}{dh} = \varphi(h) = \frac{dQ}{d(h)} \approx \frac{\Delta Q}{\Delta h}. \tag{6}$$

If we assume that ΔQ reflects no influence other than cumulative footage of drilling, then $f(h)$ may be determined as:

$$f(h) = \int_0^h f(x)dh = Q_h. \quad (7)$$

Hubbert has proposed a simple exponential equation as the function form of $\varphi(h)$:

$$\varphi(h) = Ke^{-Bh}. \quad (8)$$

It follows, then, that $f(h)$ would be of the form $f(h) = \dfrac{K - Ke^{-Bh}}{B}$, and:

$$F(\infty) = K / B = Q_\infty. \quad (9)$$

Hubbert's method of evaluating *K* and *B* was designed to assure two postulates:
1. The area under the curve determined by the mentioned function to the last (*i*th) increment of drilling employed (h_i) is equal to total quantity of a mineral resource discovered by that amount of drills;
2. The fitted curve is forced to pass through $(dQ/dh)_i$, and the rate of mineral resource discovery at h_i is equal to *i*th multiple of a distinct length of drilling.

Thus, Hubbert derived the following equation for the parameter *B*:

$$B = \frac{\ln[K / (dQ / dh)_i]}{h_i}, \quad (10)$$

where *i* = the last increment of the drilling footage. It follows that giving specified value to *K*, it would be possible to compute *B*, and then evaluate the integral of the resulting function to see if it equals Q_t:

$$Q_t) = K\int_0^h e^{-Bh} dh. \quad (11)$$

Seven years later Lieberman [26] used Hubbert's method to estimate the US ultimately recoverable reserves of uranium, e.g., a non-renewable mineral resource. However, in this case he was obliged to use a current dollar – cost of discoveries model.
Let us suppose that the original endowment[5] of a non-renewable resource were homogenous with respect to quality. Any mining of the resource and its usage in industry would result in depletion of the original endowment. Obviously, the final result of continued mining would be exhaustion of the resource. Further, in case of a non-homogenous endowment let suppose that the highest-quality resources will be mined first. In this case the degree of depletion would vary with quality classes of the endowment decreasing with the decreasing quality.
Again, let suppose that the two highest quality classes have been exhausted by time t_1 and all mining by time t_1 were of the third class. Now, suppose that the new technology and capital productivity have lowered production cost so much that at t_1 the real cost of a unit of production from the third-class resource is the same that it was at t_0 from the first quality class. Let us refer to the diminishing of our endowment of resources, which can be produced to a constant dollar cost as economic depletion. In such case a physical depletion of the endowment caused by mining may not result in economic depletion. Given that, we must analyze mineral resources at a specific cost in current dollars, and there are two possibilities of handling the problem. One way is to remove the effects of the increased productivity from our production data and inflation so that production quantities represent a constant quality of ore and to perform the analysis upon quantities of constant quality. A better way, proposed by De Verte Harris [12], however, is to adjust the data for the effects of inflation to future improvements in productivity so that the analysis is based upon quantities at constant economic costs. In this case proper adjustment for inflation should be evaluated. Let us assume that the unadjusted discovery data for a mineral non-renewable resource defines an exponential relationship perfectly. Furthermore, let us assume that inflation has caused a loss of cheap reserves at a constant rate of *x* percent per year. Given these assumptions, making an appropriate adjustment of the data for inflation prior the fitting the exponential would be equivalent to the corresponding slope of the curve with unadjusted data. Thus, the curve with adjusted data f^* could be determined as by multiplying φ, the function of unadjusted data, by another function ϕ [12]:

$$f^*(h) = \varphi(h)\phi(h), \tag{12}$$

then, let:

$$\phi(h) = e^{rt(h)}, \tag{13}$$

[5] The endowment of a region implies that quantity of a mineral in deposits meets specified physical characteristics such as quality, size and depth; see, for instance, [12].

where r = the rate of loss of reserves, expressed in decimal fraction, so that $r = x / 100$, and $t(h)$ = a fraction, which associates time with cumulative value of the drilling footage. Then, given tat φ is exponential, with a negative slope b, we have:

$$f^*(h) = e^{rt(h)} K e^{-\beta h}$$
$$f^*(h) = K e^{rt(h) - \beta h} \quad (14)$$

If $t(h) = ah$, then:

$$f^*(h) = K e^{(\alpha r - \beta) h}. \quad (15)$$

And from the equation (8)-(11) it derives that:

$$f^*(\infty) = -K / (\alpha r - \beta) = Q_\infty, \quad \alpha r < \beta \quad (16)$$

and, furthermore, Hubbert's method may be used.

Thus, from the macroeconomic point of view, a long-term forecasting and evaluation of recoverable resources may be deducted to estimation of a, b, K and r values. In reality, of course, the problem is much more complex as real geological and metallogenic framework should be taken into account. Morphology of ore bodies typical for different kinds of mineral resources and, correspondingly, different strategies of their exploration must be considered as well. In economic terms this approximation could be expressed by selection of exact value for $\tilde{h}_\infty$.

In Georgia *the premier case* economic analysis of this type has been carried out by us [49]. Figure 10 represents an economic appraisal of possible exploration results for main metallic resources of Georgia whereas Figure 11 contains the same information for industrial minerals. In both cases the adjustment to reserves is expressed in increasing their price indices. The model is based on two fundamental assumptions:

1. The ultimately recoverable investment is made in exploration of the first deposit of each resource; afterwards 25% of the income from exploitation of the first mine is reinvested in further exploration of the same resource;
2. All exploration works are started simultaneously.

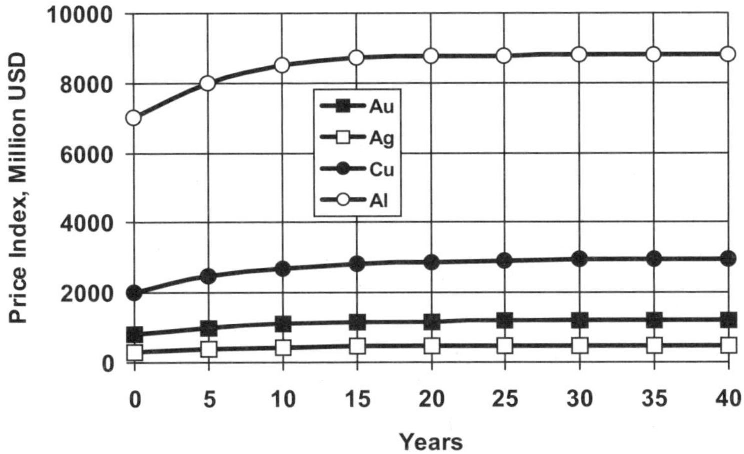

Figure 10. Economic Model of Exploration Results for Main Metallic Resources

Analysis of such curves leads to the creation of a macroeconomic model of the mineral resource base, which would describe correlation between investments and the increasing of the price indices of the ultimately recoverable resources. Figure 12 represents such a cumulative model. It may be clearly seen that this model, like all probability processes has a well-expressed extremum (Gauss distribution). Due to this feature, from the beginning efficiency of exploration is very high but afterward, in the course of increasing drilling footage, discovery probability of new reserves falls down. The model contains a margin of economically profitable investments beyond the limits of which discovered reserves could be ultimately non-recoverable.

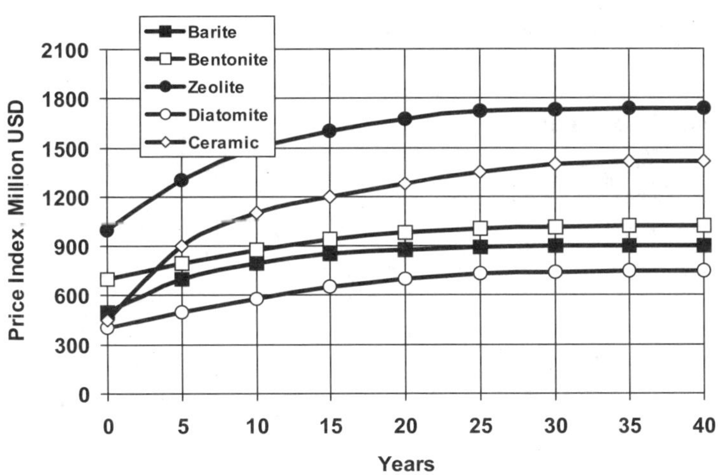

Figure 11. Economic Model of Exploration Results for Main Industrial Minerals

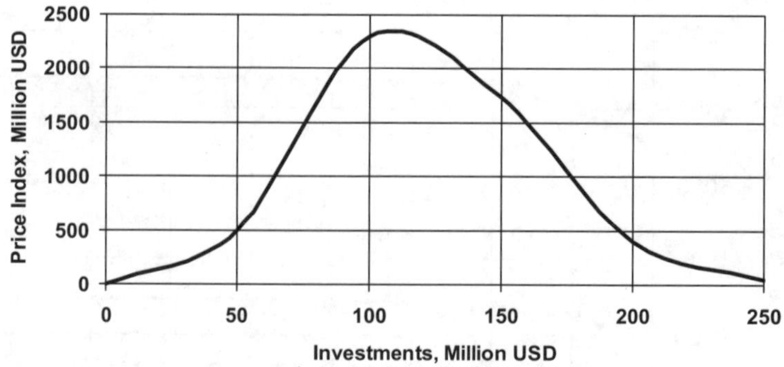

Figure 12. Macroeconomic Model of Exploration

Thus, 650 million USD investment in the Georgian mineral resource base may increase the price index of ultimately recoverable reserves by 10 billion USD. In other words, the ultimate potentiality of the Georgian mineral resource base is about one hundred billion USD.

In addition to this model, we are able to create a macroeconomic model of mining activity in Georgia. Figure 13 contains such a model, which has been drawn with respect to routine macroeconomic modelling. Of course, the model describes a possible economic effect of optimized investment dynamics. Probability of the model (evaluated by the Student's *t*-criterion) is 95%. The model contains generalized data obtained as a result of generalization of numeric expert data on different Georgian regions. Non-contradictory sense of the model consists in optimal (from the economic point of view) and simultaneous exploitation of the most profitable mineral resources. In other words, the model describes an ideal framework, exponential approach to which would determine economic efficiency of governmental management of the Georgian mineral resource base.

It could be seen that mining activity may become one of the main branches of economy, which would determine the economic revival of Georgia within the nearest 25-30 years. The sole condition for such macroeconomic framework is an adequate governmental management of the resource base.

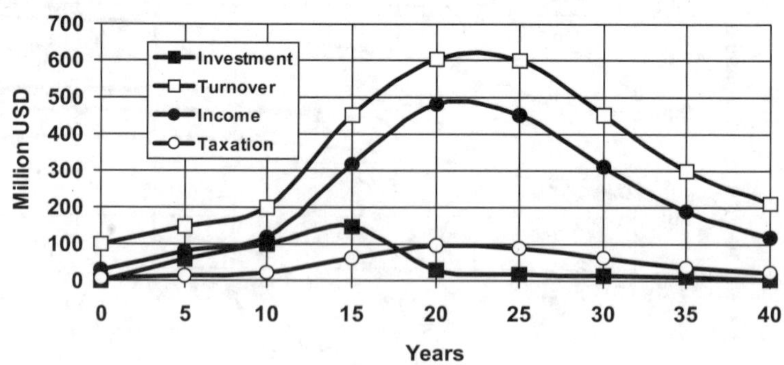

Figure 13. Macroeconomic Model of Mining Activity

8. Appendix

MAIN DEPOSITS OF MINERAL RESOURCES IN GEORGIA
(N° IN TABLES CORRESPOND TO THOSE ON MAPS)

A. *Energetical Resources*

TABLE A.1. Hydrocarbons

N°	Deposit	Enclosing Rocks	Reserves		
			Hydrocarbon	Unit of measure	Value
1	Sartichala	Oligocene sandy-clayey sequence	Oil Gas	thousand t mln m^3	13966.6 1732.438
2	Norio-Martkopi	same	Oil Gas	thousand t mln m^3	14874.3 2266.252
3	Satskhenisi	same	Oil Gas	thousand t mln m^3	1927.1 18.929
4	Mirzaani	same	Oil Gas	thousand t mln m^3	367.6 61.321
5	Patara Shiraki	same	Oil Gas	thousand t mln m^3	293.5 14.946
6	Taribani	same	Oil Gas	thousand t mln m^3	1001.70 16.996
7	Supsa	Quaternary sandy sequence	Oil Gas	thousand t mln m^3	17.0 11.703
8	Rustavi	Oligocene sandy-clayey sequence	Gas	mln m^3	5157.000

TABLE A.2. Coal/Peat

N°	Raw Material	Deposit	Technologic characteristics	Reserves	
				Unit of Measure	Value
9	Hard coal	Tkvarcheli	moisture - 1.5-3%, ash -34.7%, calorific capacity - 3.500-7.500kcal/kg	thousand t	19550
10	Hard coal	Tkibuli-Shaori	moisture - 6-10%, ash - 10-12%, calorific capacity 5970-6300 kcal/kg	thousand t	377970
11	Brown coal	Akhaltsikhe	moisture - 12-13%, ash - 36%, calorific capacity - 3230-4500 kcal/kg	thousand t	75766
12	Peat	Kobuleti	moisture - 94.3%, ash - 22%	thousand m^3	2003
13	Peat	Imnati	moisture - 91%, ash - 23%	thousand m^3	29906
14	Peat	Nabadi	moisture - 90%, ash - 70%	thousand m^3	12385
15	Peat	Anaklia	moisture - 89%, ash - 45%	thousand m^3	28074
16	Peat	Poti	moisture - 80%, ash - 31%	thousand m^3	26731

TABLE B. Geothermal Sources

N°	Source	Host Rocks	Tempe-rature, °C	Minera-lization, g/l	Main solved compounds	Interval of water presence in drills, m	Output, l/sec
17	Ohurey	Upper Cretaceous	106	1.07	Ca	2865-3300	1.07
18	Kingi	Upper Cretaceous	107	1.05	Ca	2670-3100	4.30
19	Tsaishi	Upper Cretaceous	103	1.20	Ca, Mg	700-1000	18.50
20	Zugdidi	Upper Cretaceous	105	0.85	Na, K, Ca, Mg	1800-1880	60.20
21	Samtredia	Upper Neogene	50	1.80	Na, K	1269-1280	5.20
22	Mendji	Upper Cretaceous	55	2.10	Na, K, Ca, Mg	1816-2057	2.10
23	Lisi	Palaeogene	64	0.30	Na, K	505-1245	4.0

TABLE C. Metallic Resources.

N°	Deposit	Type	Wallrock alteration	Con-stitu-ent	Grade, %	Reserves		
						item	Unit of measure	Value
1	2	3	4	5	6	7	8	9
24	Dzama	iron-skarn	skarn	Fe	32.14	ores Fe	thousand t thousand t	16667 5357
25	Poladauri	hydrothermal	quartz-sericite-chlorite	Fe	36.00	ores	thousand t thousand t	6173 2223
26	Tkibuli-Shaori	sedimentary	none	Fe	34.40	ores Fe	thousand t thousand t	210000 72240
27	Supsa-Natanebi	placer	none	Fe	2-3	ores	thousand t thousand t	770000 15400
28	Chiatura	sedimentary	none	Mn	16.6-33.4	ores Mn	million t thousand t	225.7 46688.0
29	Chkhari - Adjameti	sedimentary	none	Mn	15.8-28.2	ores Mn	million t thousand t	10.033 2367.4
30	Kvirila	sedimentary	none	Mn	19.2-23.3	ores Mn	million t thousand t	27 5307.0
31	Shkhmeri	sedimentary	none	Mn	20.7-28.9	ores Mn	million t thousand t	6398 1424.8
32	Adange	Cu-pyrrhotite hydrothermal-sedimentary metamorphosed	Quartz-sericite-chlorite	Cu	2.90	ores Cu	thousand t thousand t	8620 250

1	2	3	4	5	6	7	8	9
33	Zeskho	same	same	Cu Zn	3.50 1.67	ores Cu Zn	thousand t thousand t	22880 801 101
34	Artana	Cu-pyrrhotite vein	same	Cu	3.87	ores	thousand t thousand t	9050 350
35	Madneuli	Au-Cu-Pb-Zn-barite massive sulphide	quartz-alunite-sericite	Cu Pb Zn S Se Te Au Ag barite	1.28 0.62 4.09 6.92 0.00071 0.00076 0.00018 0.00092 36.3	ores Cu Pb Zn S Se Te Au Ag barite	thousand t thousand t thousand t thousand t thousand t t t kg t thousand t	93070 521 8 80 717 561 600 48247 100 1662
36	Sakdrisi	same	same	Au Ag Cu	0.00019 0.00030 1.03	ores Au Ag Cu	thousand t kg t thousand t	75600 19800 22 77.6
37	Tsiteli Sopeli	same	same	Au Ag Cu	0.00007 0.00040 1.58	ores Au Ag Cu	thousand t kg t thousand t	12840 8000 4 202.8
38	Kvemo Bolnisi	same	same	Cu barite	1.46 28.6	ores Cu barite	thousand t thousand t thousand t	5479 80 583
39	David Garedji	same	same	Au Ag Pb Zn Cu barite	0.00025 0.00150 2.50 5.00 0.80 19.6	ores Au Ag Pb Zn Cu barite	thousand t kg t thousand t thousand t thousand t thousand t	2000 5000 169 50 100 16.8 1948
40	Merisi group	porphyry Cu-polymetallic	quartz-sericite-chlorite	Au Ag Pb Zn Cu Bi	0.00007 0.00187 1.80 1.40 2.84 0.01	ores Au Ag Pb Zn Cu Bi	thousand t kg t thousand t thousand t thousand t t	3268 859 18.3 17.4 15.2 77.9 180.5
41	Dambludi	hydrothermal	quartz-sericite-chlorite	Au Ag Pb Zn Cu Cd In Bi	0.00019 0.00301 2.67 5.31 0.76 0.03 0.00260 0.017	ores Au Ag Pb Zn Cu Cd In Bi	thousand t kg t thousand t thousand t thousand t t t t	1869 1882 56 48 96 13.7 555 21 181

1	2	3	4	5	6	7	8	9
42	Kvaisi	hydrothermal	quartz-carbonate-kaolinite	Pb Zn Ag Cd	2.24 6.47 0.00154 0.02	ores Pb Zn Ag Cd	thousand t thousand t thousand t t t	2851 56.2 197.7 18 279
43	Skatykom	same	same	Pb	6.81	ores Pb	thousand t thousand t	174.5 9.8
44	Razdaran-kom	same	same	Pb Zn	1.50 1.00	ores Pb Zn	thousand t thousand t thousand t	340 5.1 0.99
45	Rtskhme-luri	same	silicification	Pb Zn	1.61 2.67	ores Pb Zn	thousand t thousand t thousand t	255 4.1 6.8
46	Ertso	same	quartz-carbonate-kaolinite					
47	Amtkeli	same	kaolinization	Pb Zn	2.36 3.80	ores Pb Zn	thousand t thousand t thousand t	285 3.2 7.3
48	Brdzyshra	strata-bound	carbonate-sericite	Pb Zn	1.20 1.00	ores Pb Zn	thousand t thousand t thousand t	250 15.5 23.2
49	Dzyshra	same	same					
50	Karobi	plutonogenic hydrothermal	silicification	Mo	0.98	ores Mo	thousand t t	55 50
51	Notsata	same	same	W Au	3.58 0.00023	ores W Au	t t kg	10826 50 1882
52	Tsana	hydrothermal	quartz-sericite	As	16.4	ores As	thousand t thousand t	355 55
53	Chorokhi	same	same	As Sn	12.71 0.14	ores As Sn	thousand t thousand t t	39.4 4.5 41
54	Lukhumi	telethermal	calcitization	As Sb Au	7.80 7.32 0.00013	ores As Sb Au	thousand t thousand t thousand t kg	150 11.1 1.8 1404
55	Akhey	telethermal	dickitization	Hg	0.34	ores Hg	thousand t t	824 2546
56	Avadkhara	same	same	Hg	0.29	ores Hg	thousand t t	480 1353
57	Akhakhcha	same	same	Hg	0.50	ores Hg	thousand t t	440 2200
58	Talakhiani	same	same	Hg	0.41	ores Hg	thousand t t	75 300

1	2	3	4	5	6	7	8	9
59	Gomi	telethermal	dickitization	Hg	0.27	ores Hg	thousand t t	98.78 289
60	Zopkhito	hydrothermal	quartz-sericite-chlorite	Sb Au Ag	12.39 0.00030 0.00137	ores Sb Au Ag	thousand t thousand t kg t	231.4 27.418 8789 39
61	Chveluri	same	same	Sb	10.81	ores Sb	thousand t thousand t	121 12.746
62	Gelati	sedimentary	none	Al_2O_3	18.07	ores Al	thousand t thousand t	176198 16856
63	Enguri basin	placer	none			Au	kg	>3000
64	Khrami basin	same	same			Au	kg	>4000

TABLE D. Chemical & Agrochemical Resources

N°	Deposit	Type	Host rocks	Technologic indices	Reserves	
					Unit of measure	Value
1	2	3	4	5	6	7
65	Apshra	barite	Late Jurassic limestones	$BaSO_4$ grade - 45.6%	thousand t of barite	2648
66	Amakekva	same	Early Jurassic schist	$BaSO_4$ grade - 46.0%	thousand t of barite	98
67	Adzaga	same	Late Jurassic limestones	$BaSO_4$ grade - 39.8%	thousand t of barite	845
68	Chordi	same	Middle Jurassic volcanic rocks	$BaSO_4$ grade - 73%	thousand t of barite	1862
69	Pitikvara	same	same	$BaSO_4$ grade - 95%	thousand t of barite	130
70	Kudaro	same	same	$BaSO_4$ grade - 87%	thousand t of barite	275
71	Khaishi group	same	same	BaSO4 grade - 80%	thousand t of barite	306
72	Goumbri	bentonite	Early Cretaceous trachites	Content (%): SiO_2 - 62.08, Al_2O_3 - 12.30, $Na_2O + K_2O$ - 0.58, Na_2O - 11.40	thousand t	4100
73	Askana	same		Content (%): SiO_2 - 62.08, Al_2O_3 - 12.30, $Na_2O + K_2O$ - 0.58, Na_2O - 11.40	thousand t	7518
74	Chorchana group	talc	Palaeozoic hyperbasites	talc content - 50-60%	thousand t of talc	2488

1	2	3	4	5	6	7
75	Dzegvi	zeolite	Neogene volcanic-sedimentary rocks	clinoptillolite content - 70%, mordenite content - 10%	thousand t	1735
76	Tedzami	same	same	clinoptillolite content - 66%, mordenite content - 11%	thousand t	30381
77	Kisatibi	diatomite	Neogene lacustrine volcanic rocks	SiO_2 - 84-95%, free SiO_2 - 3.5-11.1%, Fe_2O_3 - 0.3-1.5%, density - 0.56-0.83 g/cm^3	thousand t	7808
78	Paravani	perlite	Neogene volcanic flow	distending interval - 1000-1200°C, distension coefficient - 5-14, volumetrical weight of the distended mass - 150- 400 kgf/cm^3	thousand m^3	60515
79	Tsikhisjvari	Acid-resisting andesite	Quaternary volcanic flow		thousand m^3	>5000
80	Gldani	Glauber salt	Quaternary evaporites	Na_2SO_4 content in dry mirabillite - 40.86%	thousand t of mirabillite	65
81	Grdzeli Tba	same	modern brine lake	Na_2SO_4 content in dry mirabillite - 84.95%	thousand t of mirabillite	608
82	Sakhare Tba	same	same	Na_2SO_4 content in dry mirabillite - 86.13%	thousand t of mirabillite	385
83	Kutaisi	phosph-orites	Oligocene	P_2O_5 grade - 8.0%	thousand t	250
84	Lechkhumi	same	same	P_2O_5 grade - 10.8%	thousand t	407
85	Algeti	lithogra-phic stones	Late Cretaceous limestones	ultimate strength on compression - 1300 kg/cm^2, ultimate strength on bending - 270 kg/cm^2, porosity - 4%, diamond yield - 5%	thousand m^3	1620
86	Taribani	haloids	petroleum waters	Content (mg/eq): I - 20.3, Br - 14.0, B_2CO_3 - 23.5		
87	Adjameti	chalce-dony	Oligocene clayey rocks	SiO_2 content in chalcedony - 98.56, refractory capacity - 1750°C, density - 2.62 g/cm^3	thousand t	4090
88	Brdzysh-kha	mineral paints	Quaternary weathering	Fe_2O_3 content - 61.82%, pigment - ochre, oil capacity - 26.00%	thousand t	600
89	Chititskaro	same	same	Fe_2O_3 content - 74.90%, pigment - limonite, oil capacity - 31.88%	thousand t	350
90	Matkhoji-Odzlouri	same	same	Fe_2O_3 content - 22.65%, pigment - limonite, oil capacity - 30.55%	thousand t	437
91	Shrosha-Ubisi	same	same	Fe_2O_3 content - 67.98%, pigment - ochre, oil capacity - 23.00%	thousand t	250
92	Agara	same	same	Fe_2O_3 content - 21.20%, pigment - limonite, oil capacity - 30.00%	thousand t	1600

TABLE E. Ceramic Materials

N°	Deposit	Type	Host rocks	Technologic indices	Reserves	
					Unit of measure	Value
93	Shrosha	pottery pegmatite	Palaeozoic granites	Content (%): feldspar - 30-59, quartz - 18-50.	thousand t	2232
94	Jvarisi	porcelain kaolinite	Middle Juassic clays	$Fe_2O_3 > 2.5\%$, $Al_2O_3 < 25\%$	thousand t	1298

TABLE F. Gems

N°	Deposit	Type	Host rocks	Reserves	
				Unit of measure	Value
95	Akhaltsikhe	agate	quartz veins in Neogene volcanics	million kg	2
96	Dmanisi	same	Same	million kg	1
97	Goderdzi	opal	opalized Neogene wood	unlimited	
98	Kayundag	obsidian	Neogene ryolite flow	unlimited	

TABLE G. Facing Materials

N°	Deposit	Type	Host rocks	Technologic indice	Reserves	
					Unit of measure	Value
1	2	3	4	5	6	7
99	Rikoti	gabbro-"rikotite"	Palaeozoic gabbro-monzonites	20.5	46	4918
100	Tsipi	quartz diorite	Palaeozoic quartz diorites	37.8	11.2	802
101	Rkvia	granite	Jurassic granite	for crumb		20000
102	Kursebi	teschenite	Middle Jurassic dykes	20	6	4556
103	Opurchkheti	same	same	15.2	9.65	3037
104	Aigbi	gabbro-diabase	Late Jurassic dyke	17.2	15.1	32490
105	Chirkvali	same	Early Jurassic dyke	15.16	9.98	1515
106	Kvaishauri	same	Eocene dyke	14.7	10.0	1147
107	Ushlevi	diabase	Middle Jurassic dyke	17.0	10.6	740
108	Tsalka	basalt	Tertiary dolerite flows	15.2	10.0	3173

1	2	3	4	5	6	7
109	Beshtasheni	same	same	36.48	10.40	3046
110	Kazreti	same	same	for crumb		326
111	Ratevani	same	same	24.1	10.0	4523
112	Marneuli	same	same	40	20	6825
113	Karadag	same	Tertiary volcano flow	for crumb		1212
114	Shemokmedi	andesite	Eocene andesites	25.12	12.31	741
115	Kobisi	andesito-dacite	Quaternary flow	25.2	7.2	5888
116	Bolnisi group	decorative tuff	Late Cretaceous tuffs	25.2	12.1	938
117	Chivchavi-I	same	same	60.6	11.9	25484
118	Chivchavi-II	same	same	39.7	12.9	1509
119	Dizi	marble	Triassic marbles	20.5	12.0	568
120	Chuberi	same	same	22.42	9.22	5444
121	Choluri	same	same	20.32	10.20	1740
122	Kldiana	marmorized limestones	Late Cretaceous	61.6	12.0	9288
123	Gumista	same	same	for crumb		6986
124	Shroma	same	same	59.10	15.00	10005
125	Salieti	same	Early Jurassic	17.7	6.83	23936
126	Moliti	same	same	7.75	4.26	4512
127	Dzveli Shrosha	same	same	17.2	9.3	150
128	Akhali Shrosha	same	same	17.7	8.9	231
129	Merelisi	same	same	for crumb		212
130	Chobareti	same	Late Cretaceous	11.21	5.5	26680
131	Kanchaveti	same	Neogene	for crumb		875
132	Jakhuri	same	Late Cretaceous	70.7	10.2	583
133	Lopota	same	Early Jurassic	14	10	1443
134	Ilto	same	Late Cretaceous	for crumb		740
135	Sadakhlo	same	same	12.7	9.7	1881
136	Eklari	limestone	same	45	10	4438
137	Darkveti	same	same	86.8	23	6115
138	Kvakhchiri	same	same	85.2	14.7	1065
139	Motsameta	same	same	75.2	10	18967

TABLE H. Inert, Metallurgic & Building Resources

N°	Deposit	Type	Host rocks	Technologic indices	Reserves: Unit of measure	Value
1	2	3	4	5	6	7
140	Sadakhlo	Flux limestone	Late Cretaceous	density - 2.63 g/cm^3, ultimate strength - 1400 kgf/ cm^2	thousand t	19648
141	Dedoplis-tskaro	same	Neogene	density - 2.62 g/cm^3, ultimate strength - 1032 kgf/cm^2	thousand t	50864
142	Chishura	same	Early Cretaceous	density - 2.62 g/cm^3, ultimate strength - 1032 kgf/cm^2	thousand t	12723
143	Chkheishi	gravel & sand	Modern alluvial	granulometry (mm/%): <5 - 21.1, 5-10 - 6.0, 10-20 - 11.4, 20-4 - 19.6, 40-70 - 15.3, >70 - 28.3	thousand m^3	11541
144	Bebnis	same	Same	granulometry (mm/%): <5 - 27, 5-10 - 11, 10-20 - 12, 20-4 - 12, 40-70 - 17, >70 - 18	thousand m^3	10685
145	Enguri-I	same	same	granulometry (mm/%): <5 - 20.2, 5-10 - 12.4, 10-20 - 9.1, 20-40 - 11.5, 40-70 - 16.5, >70 - 28.1	thousand m^3	36373
146	Enguri-II	same	same	granulometry (mm/%): <5 - 26.2, 5-10 - 2.9, 10-20 - 14.2, 20-40 - 13.2, 40-70 - 22.4, >70 - 5.1	thousand m^3	10419
147	Tsageri	same	same	granulometry (mm/%): <5 - 28.2, 5-10 - 9.7, 10-20 - 10.4, 20-40 - 14.2, 40-70 - 20.2, >70 - 22.2	thousand m^3	13987
148	Kavtiskhevi	same	same	granulometry (mm/%): <5 - 25.4, 5-10 - 15.0, 10-20 - 14.8, 20-40 - 14.6, 40-70 - 12.2, >70 - 16.4	thousand m^3	6614
149	Sartichala	same	same	granulometry (mm/%): <5 - 37.2, 5-10 - 6.3, 10-20 - 10.6, 20-40 - 9.8, 40 -70 - 20.3, >70 - 18.8	thousand m^3	10906
150	Khrami	same	same	granulometry (mm/%): <5 50.0, 5-10 - 6.8, 10-20 - 6.2, 20-40 - 17.5, 40-70 - 5.0, >70 - 4.7	thousand m^3	42609
151	Badjini	sand for glass	Miocene sea littoral sands	refractory capacity - 1610°C, quartz - 94%, feldspar - 5.4%	thousand t	9078
152	Itkhvisi	same	Oligocene sea sands	refractory capacity - 1600°C, quartz - 80%, feldspar - 15%	thousand t	1472
153	Saparis Gele	same	same	refractory capacity - 1700°C, quartz - 66%, feldspar - 24%	thousand t	531
154	Shukruti	same	same	refractory capacity - 1630°C, quartz - 85%, feldspar - 7%	thousand t	1410

1	2	3	4	5	6	7
155	Surami	foundry sands	Miocene littoral sands	quartz - 75-85%, feldspar - 2-15%, mica - 9-11%	thousand t	2825
156	Kroli	same	Oligocene sea sands	quartz - 76-98%, feldspar - 3-21%, mica - 3%	thousand t	3408
157	Kldis Tskaro	same	same	quartz - 29-54%, feldspar - 1-35%, mica - 0-3%	thousand t	7059
158	Itavazi	same	same	quartz - 20-88%, feldspar - 1-45%, mica - 0-8%	thousand t	52451
159	Gali	chalk	Late Cretaceous	$CaCO_3$ - 95.87 - 98.87%	thousand t	2928
160	Okumi	same	same	$CaCO_3$ - 97.57%	thousand t	3962
161	Otkhari	limestones for lime	Late Cretaceous	active $CaCO_3$+ $MgCO_3$ - 75-86%, slaking - 1000°C, yield - 2.90l/kg, unslaked particles - 10%, slaking rate - 4 min	thousand t	8467
162	Skuri	Same	same	active $CaCO_3$ + $MgCO_3$ - 88-90%, slaking - 1150°C, yield - 2.65l/kg, unslaked particles - 0.18%, slaking rate - 13 min	thousand t	6743
163	Ambrolauri	same	same	active $CaCO_3$+ $MgCO_3$ - 87%, slaking - 1200°C, yield - 2.41l/kg, unslaked particles - 1.67%, slaking rate - 12 min	thousand t	3156
164	Tetri Tskaro	same	same	active $CaCO_3$+ $MgCO_3$ - 85%, slaking - 1200°C, yield - 3.601l/kg, unslaked particles - 4%, slaking rate - 12 min	thousand t	34490
165	Iori	limestones for lime	Neogene	active $CaCO_3$+ $MgCO_3$ - 72%, slaking - 1200°C, yield - 2.301l/kg, unslaked particles - 9%, slaking rate - 25 min	thousand t	3109
166	Dedoplis-tskaro	same	same	active $CaCO_3$ + $MgCO_3$ - 95%, slaking - 1250°C, yield - 3.601l/kg, unslaked particles - 3%, slaking rate - 7 min	thousand t	10700
167	Tkvarcheli	metallurgical dolomite	Early Cretaceous	refractory capacity - 1760°C	thousand t	41444
168	Jgali	same	same	refractory capacity - 1760°C	thousand t	5757
169	Abano	same	same	refractory capacity - 1750°C	thousand t	5029
170	Saskhori	limestone for cement	Late Cretaceous	content (%): SiO_2 - 10.25, CaO - 47.78	thousand t	80293

1	2	3	4	5	6	7
171	Kaspi	same	Neogene	content (%): SiO_2 - 5.61, CaO - 53.70	thousand t	30493
172	Gardabani	clays for cement	Quaternary alluvial	SiO_2/R_2O_3+RO - 2.26 Al_2O_3/R_2O_3+RO - 2.60, density - 1.60-1.90 g/cm^3	thousand t	4728
173	Tkvarcheli	refractory clays	Middle Jurassic	melting - 1710°C, shrinkage - 1.65%, water adsorption - 8.55%	thousand t	3325
174	Tsetskhleuri	same	Neogene	melting - 1720°C, shrinkage - 5.00%, water adsorption - 7.00%	thousand t	1642
175	Chartali	same	Middle Jurassic	melting - 1700°C, shrinkage - 3.00%, water adsorption - 10.00%	thousand t	2000
176	Tkibuli	same	same	melting - 1750°C, shrinkage - 7.00%, water adsorption - 10.98%	thousand t	5000
177	Shrosha	refractory clays	Early Jurassic	melting - 1710°C, shrinkage - 8.30%, water adsorption - 12.00%	thousand t	2558
178	Volodari	brick earth	Quaternary alluvial	refractory capacity - 1200°C, foundry moisture - 27.28%, shrinkage - 14.0%, water adsorption - 22.00%	thousand m^3	1977
179	Achigvara	same	same	refractory capacity - 1445°C, oundry moisture - 41.00%, shrinkage - 15.4%, water adsorption - 15.60%	thousand m$_3$	4063
180	Lusuliavi	same	same	refractory capacity -1200°C, foundry moisture - 23.66%, shrinkage - 8.9%, water adsorption - 18.97%	thousand m^3	1126
181	Naesakao	same	same	refractory capacity - 1320°C, foundry moisture - 38.00%, shrinkage - 9.8%, water adsorption -24.09%	thousand m^3	4161
182	Sagaredjo	same	same	refractory capacity -1140°C, foundry moisture - 22.85%, shrinkage - 9.4%	thousand m^3	2040
183	Miriani	same	same	refractory capacity - 1250°C, foundry moisture - 21.00%, shrinkage - 9.00%, water adsorption - 25.89%	thousand m^3	15550
184	Alazani	same	same	refractory capacity - 1190°C, foundry moisture - 23.22%, shrinkage - 9.00%, water adsorption - 4.80%	thousand m^3	1379
185	Dedoplis-tskaro	same	same	refractory capacity - 1150°C, foundry moisture - 25.20%, shrinkage - 20.60%, water adsorption - 5.60%	thousand m^3	6437

1	2	3	4	5	6	7
186	Intsoba	roof shales	Early Jurassic schists	yield of shale blocks - 33.7%, yield of plates - 40.7%	thousand m^3	4206
187	Khudoni	gypsum	Late Jurassic evaporites	$CaSO_4.2H_2O$ - 88.70%	thousand t	23898
188	Tskaltbila	same	Neogene evaporites	$CaSO_4.2H_2O$ - 98.30%	thousand t	3383
189	Mukhli-Tsesi	same	Late Jurassic evaporites	$CaSO_4.2H_2O$ - 99.45%	thousand t	3750
190	Navtlugi	anhydrite	Quaternary brine sediments	$CaSO_4$. - 40.75%	thousand t	705
191	Marneuli	same	Neogene evaporites	$CaSO_4$. - 46.16%	thousand t	1090
192	Taribani	same	Quaternary brine sediments	$CaSO_4$. - 38.94%	thousand t	807

9. References

1. Adamia, Sh.A. (1974) *Pre-Jurassic formations of the Caucasus*, Metsniereba Publishers, Tbilisi, p. 354 (in Russian).
2. Adamia, Sh.A. (1977) *Tectonics and geological history of Abkhazia,* Metsniereba Publishers, Tbilisi, p. 67 (in Russian).
3. Adamia, Sh.A., Buadze, V.I., and Shavishvili, I.D. (1977) Caucasus in the Phanerozoic: a geodynamic model, in S. Jankovic (ed.), *Metallogeny and Plate Tectonics in the North-eastern Mediterranean,* Belgrade, 215-230.
4. Adamia, Sh.A., Zakariadze, G.S., and Lordkipanidze, M.B. (1977) Evolution of an ancient continental margin as an example of the Alpine history of Caucasus, *Geotectonics* **4**, 88-103 (in Russian).
5. Adamia, Sh.A., Chkhotua, T., Kekelia, M., Lordkipanidze, M., and Zakariadze, G. (1981) Tectonics of the Caucasus and adjoining regions: implications for the evolution of the Tethys ocean, *Jour. Struct. Geol.* **3**, 437-447.
6. Adamia, Sh., Belov, A., Chabukiani, A., Chkhotua, T.G., Lordkipanidze, M.B., and Shavishvili, I. (1987) Terrane description: the Caucasus, *Annales Géologiques des Pays Helléniques* **37**, 537-560.
7. Baakashvili, G.T. (1970) *The mercury belt of Abkhazia,* Georgian Institute for Scientific & Technical Information, Tbilisi, p. 46 (in Russian).
8. Khitarov, N.I. (ed.) (1981) *Barite 1981*, Nauka Publishers, Moscow, p. 253 (in Russian).
9. Beridze, M.A. (1973) Early Jurassic spilite-keratophyre volcanism of the North-Western Abkhazia and its influence on clastic sediment formation, in A. Tsagareli (ed.), *Problems of Geochemistry and Petrology,* Metsniereba Publishers, Tbilisi, 22-56 (in Russian).
10. Borsuk, A.M. (1977) Lateral variations of composition of the Jurassic magmatic formations of the Greater Caucasus and their petrochemical meanings, in F. Chukhrov (ed.), *Magmatic Formations of the Caucasus and East-European Platform*, Nauka Publishers, Moscow, 87-126 (in Russian).
11. Buadze, V.I., Tvalchrelidze, A.G. (1980) Geological, geochemical, and physico-chemical conditions of formation of stratiform non-ferrous deposits in sedimentary rocks of the Caucasus, in J.D. Ridge (ed.), *Proc. Fifth Quadrennial IAGOD Symposium,* E. Schweizerbart'sche Verlagbuchhandlung, Stuttgart, **1,** 245-260.
12. De Verte Harris, P. (1984) *Mineral resources appraisal*, Clarendon Press, London, p. 445.

13. Dzotsenidze, G.S. (1948) *Pre-Miocene effusive magmatism of Georgia*, Tbilisi, Georg. Acad. Sci. Publ., p. 407 (in Russian).
14. Dzotsenidze, G.S. (1969) *Role of volcanism in sedimentary rocks and ores formation*, Nedra Publishers, Moscow, p. 364 (in Russian).
15. Gamkrelidze, I.P. (1969) Structure and development of the western part of the Greater Caucasus southern slope and the Georgian Block. *Geotectonics* **4**, 72-84 (in Russian).
16. Gamkrelidze, I.P. (1997) Terranes of the Caucasus and adjacent areas, *Geor. Acad. Sci. Bull.* **155**, 391-394.
17. Gamkrelidze, P.D. (1957) *Main features of the geological structure of Georgia*, Georgian Geological Institute, Tbilisi, p. 187 (in Russian).
18. Azhgireiy, G.D. (ed.) (1957) *Geological structure of the Greater Caucasus,* Nedra Publishers, Moscow, p. 263 (in Russian).
19. Sidorenko, A.D. (ed.) (1964) *Geology of the USSR, Vol. X, Georgian SSR*, Gosgeoltekhizdat Publishers, Moscow, p. 543.
20. Gudjedjiani, B.I. *et al.* (eds.) (1974) *Geology of the USSR, Vol. X, Georgian SSR (Mineral resources)* Nedra Publishers, Moscow, p. 307 (in Russian).
21. Soc. Min. Geol. Japan (1974) *Geology of Kuroko deposits*, p. 438.
22. Gogishvili, V.G., Guniava, V.D., Ratman, I.P., and Gogishvili, T.Sh. (1976) Poste-Eocene ore formation in the Transcaucasia (Somkhito-Karabakh and Gagra-Java zones), *USSR Acad. Sci. Trans., Ser. geol.* **11,** 99-115 (in Russian).
23. Gomelauri, A.I., (1978) Formation temperatures of the Georgian barite deposits and the autoclave tests, in, G.B. Naumov (ed.), *Termobarogeochemistry of the Earth's Crust and Ore Formation,* Nauka Publishers, Moscow, 127-130 (in Russian).
24. Howell, D. (1989) *Tectonics of suspect terranes,* Chapman & Hall, p. 232.
25. Hubbert, M.K. (1969) Energy resources, in Resources and Man (A Study and Recommendations by the Committee on Resources and Man, National Academy of Sciences – National Council), W.H. Freeman & Co., San Francisco, p. 157-242.
26. Janelidze, T.V., Tvalchrelidze, A.G., Paylodze, Z.A., and Tatishvili, M.G. (1981) Jurassic volcanism of the Greater Caucasus and ore formation, *Volcanology and Seismology* **2**, 14-25 (in Russian).
27. Kekeliya, S.A., Tvalchrelidze, A.G., and Yaroshevich, V.Z. (1984) The geological and physicochemical conditions of formation of massive-sulphide-barite-base metal deposits, *Intern. Geol. Rev.* **26**, 1437-1444.
28. Khamkhadze, N.I. (1982) Oligocene tectonic hydrothermal activity of districts with siliceous-manganese sediments in Georgia, *VI IAGOD Symposium, Collected Abstracts,* Tbilisi, 292-293.
29. Khamkhadze, N.I. (1984) On the relationship of silica and ore formation in manganese deposits in Georgia, USSR, in G. Morvai (ed.), *Geology and Geochemistry of Manganese*, Hungarian Acad. Sci. Publ., Budapest, 427-431.
30. Lieberman, N.A. (1976) United States uranium resources – an analysis of historical data, *Science* **192 (4238)**, 431-436.
31. Lordkipanidze, M.B. (1980) *Alpine volcanism and geodynamics of the central segment of the Mediterranean folded belt,* Metsniereba Publishers, Tbilisi, p.162 (in Russian).
32. Machabeli, G.A. (1961) *Composition and formation conditions of bentonite clays in the Akhaltsikhe Trough. Doctoral thesis,* A. Janelidze Geological Institute, USSR, p. 325 (in Russian).
33. Machabeli, G.A. (1982) Peculiarities of sedimentogenesis and diagenesis of Oligocene manganese bearing rocks in Georgia, *VI IAGOD Symposium, Collected Abstracts,* Tbilisi, 297-298.
34. Magalashvili, G.A. (1984) Non-metallic ore deposits of Georgia, *IGC, XXVII Session, Guidebook of Excursions,* Moscow, 200-221.
35. Magalashvili G.A., Megrelishvili, G.Sh. (1989) Porcelain stones of the Caucasus, in *II World Congress on Non-Metallic Materials, Beijing (China)*, Intern. Acad. Publ., **1,** pp. 317-321.
36. Margolin, A.M. (1974) *Evaluation of mineral deposits. Mathematical methods,* Nedra Publishers, Moscow, p. 261 (in Russian).
37. Tvalchrelidze, G.A. (ed.) (1982) *Ore deposits of the Caucasus*, *Guidebook,* Metsniereba Publichers, Tbilisi, p. 141.

38. Singer, D.A., De Yung Jr, J.H. (1980) What can grade-tonnage relations really tell us? in C. Gillemin and Ph. Lagny (eds.), *Résource minérales - Mineral Resources., 26th IGC,* Orléans: BRGM, **106**, pp. 91-101.
39. Smirmov, V.I. (1982) *Geology of ore deposits*, Moscow, Nedra Publishers, p. 875.
40. Tvalchrelidze, G.A. (ed.) (1976) *Tectonics and metallogeny of the Caucasus*, Metsniereba Publishers, Tbilisi, p. 323 (in Russian).
41. Tvalchrelidze, A.A. (1941) A brief essay on the fuller's earth deposits in the Georgian SSR, in A.A. Tvalchrelidze (ed.), *Bentonite Clays of the Georgian SSR*, Tekhnika da Shroma Publishers, Tbilisi, pp. 11-46 (in Russian).
42. Tvalchrelidze, A.G. (1984) Petrologic mechanisms of origination of pyrite-bearing hydrothermal systems, *Geologica Carpathica* **35**, 91-103.
43. Tvalchrelidze, A.G. (1986) Physico-chemical conditions of base Metal massive sulphide formation, in C.H. Fridrich *et al.* (eds.), *Geology and Metallogeny of Copper Deposits,* Springer Verlag, pp. 358-369.
44. Tvalchrelidze, A.G. (1987) *Geochemical conditions of formation of base metal massive sulphide deposits,* Nedra Publlishers, Moscow, p. 188 (in Russian).
45. Tvalchrelidze, A.G. (1989) A genetic model for vein polymetallic and barite deposits in V.I. Smirnov (ed.), *Ore-Forming Processes and Systems,* Nauka Publishers, Moscow, pp. 102-112 (in Russian).
46. Tvalchrelidze, A.G. (1990) Three-dimensional quantitative models of vein-type ore deposits and a theory of rhythmical zoning, *8th IAGOD Symposium, Abstracts, A231.*
47. Tvalchrelidze, A.G. (1992) A zonal and paleotemperature model for plutonogenic polymetallic ore districts and fields, *Geology of Ore Deposits*, **1,** 58-69 (in Russian).
48. Tvalchrelidze, A.G. (1993) *Mineral Resources of Georgia*, Parliamentary Research Service, Tbilisi, Tekhinformy Publishers, p. 18.
49. Tvalchrelidze, A.G. (1995) *Development of a geological-economic system for governmental management of Georgian mineral resources,* John D. & Catherine T. MacArthur Foundation, Tbilisi, p. 89.
50. Tvalchrelidze A.G., (1998) Solid mineral resources of Georgia, *Centre for Strategic Investigations and Development, a Bulletin* **9**, 54.
51. Tvalchrelidze, A.G., Chichinadze, M.K. (1985) Features of pyrrhotite in pyritic stratiform and vein ores in black schists, *Intern. Geol. Rev.*, **27**, 492-501. **(J)**
52. Tvalchrelidze, A.G., Pataridze D.V. (1986) Experimental investigation of pyrite thermal decomposition in vacuum, *Trans. of the USSR Mineralogical Society* **CXV**, issue 6, 727-735 (in Russian).
53. Tvalchrelidze, A.G., Magalashvili, A.G., and Kikava A.A. (1991) *Thermophysical simulation of the Merisi Mining District,* A. Jandelidze Geological Institute, Tbilisi, p. 23 (in Russian).
54. Tvalchrelidze, G.A. (1961) *Endogenous metallogeny of Georgia,* Gosgeoltekhizdat Publishers, Moscow, p. 344 (in Russian).
55. Tvalchrelidze, G.A. (1966) *Systematics of endogenous deposits of the folded belts,* Nedra Publishers, Moscow, p. 235 (in Russian).
56. Tvalchrelidze, G.A. (1972) *Ore provinces of the World,* Nedra Publishers, Moscow, p. 344 (in Russian).
57. Tvalchrelidze, G.A. (1977) *Metallogenic features of the main types of volcanic belts,* Nedra Publishers, Moscow, p. 110 (in Russian).
58. Tvalchrelidze, G.A. (1980) Copper metallogeny of the Caucasus, in S. Jankovic (ed.), *European Copper Deposits,* Belgrade, pp. 191-196.
59. Tvalchrelidze, G.A. (1984) Main features of metallogeny of the Caucasus, in T.V. Janelidze and A.G. Tvalchrelidze (eds.), *Proceedings of the Sixth Quadrennial IAGOD Symposium,* E. Schweizerbart'sche Verlagbuchhandlung, Stuttgart, **1**, pp. 1-5.
60. Yaroshevich, V.Z., Tvalchrelidze, A.G. (1981) Distribution of sulphur isotopes in iron sulphides of some types of pyrite deposits, *Intern. Geol. Rev.* **23**, 524-534.
61. Zhabin, A.G., Tamazashvili, N.A., Meshveldishvili, T.A., Naniev, M.I., and Margiev, M.A. (1982) Deep forecasting of polymetallic ores at the Kvaisi ore field, *USSR Acad. Sci. Bull.* **265**, 410-413.

MINERAL RESOURCES OF RUSSIA IN THE XXI CENTURY

D.V. RUNDQVIST
Russian Academy of Sciences

1. Introduction

Mineral resources of Russia in the past history and during the last decade of the restructuring of the State, as well as in the nearest future — up to the years 2010-2015, will remain the main source of the development of the country's economy and a determining factor for the welfare of the people.

In connection with the distinct tendency towards the globalization of the world economy, the mineral resources of Russia are increasingly regarded in the context of world-wide problems of supplying raw materials for the industry, transport and social sector of both developed and developing countries. And in this regard the real state of affairs with the utilizaton of Russian raw material resources is a factor of vivid importance, determining long-term prospects of the development of the world economy, first of all — for Western Europe, which is inseparably tied with Russia by the network of petroleum pipe-lines, gas conduits and electric power lines.

2. General information

Russia is a country, rich in mineral resources. To 2.6% of population there fall 12.5% of continental firm-land, 30% of shelf water areas — the main sources of future energy resources, 22% of forest resources, 20% of fresh waters, and 16% of all kinds of mineral resources. In the balance of Russia there are more than 20 thousand mining fields (besides underground waters and construction materials), half of which are in the process of exploitation. Russia is rich in:

- Energy resources (32% - gas, 12-13% oil, 12% - coal of the world's explored reserves).
- Noble metals (platinum - 40%, palladium - 90%).
- Rare and rare-earth elements (niobium ~ 35%, tantalum ~ 80%, yttrium - 50%, lithium - 28%, beryllium - 15%, zirconium - 12%).
- Metallurgical production metals (Ni - 36%, Fe - 27%, Sn - 27%, Co - 20%, Zn - 16%, Pb - 12%).
- Agrochemical ores (potassic salts – the first place in the world, apatite and phosphorites - the second place in the world).

A.G. Tvalchrelidze and G. Morizot (eds.),
Mineral Resource Base of the Southern Caucasus and Systems for its Management in the XXI Century, 71–80.

By the explored reserves of diamonds Russia is in the first place, and of gold - in the third place in the world.
At the same time, after the collapse of the Soviet Union Russia lost reserves of a number of important kinds of mineral resources and became dependent on the import of **chromium, manganese, bauxites, bentonite (for the purification of oils), kaolin.** Furthermore, because of the loss of a number of processing plants, which happened to belong to other countries of the CIS, and because of the absence of technologies, Russia imports **titanium,** although by titanium reserves it occupies the second place in the world. Russia also imports **rare metals**, namely **zirconium**, while having considerable subsoil reserves of the mineral.
Today Russia is the largest exporter of mineral raw materials and the products of their processing. In the years 1999-2000 export of petroleum amounted to 57.3% of the output, of gas - to 32%, of coal - to 12%, of iron ore - to 14.9%, of copper – to 85%, of nickel - to 91%, of zinc - to 59%, of lead - to 9.1%, of wolfram trioxide - to 96%. Moreover, at the expense of using previously accumulated reserves, export of uranium amounted to 416.6% of output (according to the data for 1996), and that of molybdenum - to 356.2%. There is a characteristic situation with alumina. Russia exports the mineral from Ukraine and Kazakhstan, being at the same time one of the world's main exporters of aluminium.

3. Characteristic of mineral resources of the Russian territory

Throughout the whole period of the restructuring of the country's economy during the years 1992-2000, mineral resource complex of Russia accounted for the greatest part (more than 50%) of the Federal budget revenues and circa 65-70% of foreign exchange earnings. And the leading role under the circumstances was played by the fuel and energy complex — production, refining, and transportation of oil and gas.
According to the estimates by the Ministry of Natural Resources, Russia produces per year mineral resources at the cost of circa $100 billion. This is less than in the USA (the production is estimated to amount to $150 billion), approximately the same as in China, and more than in other countries.
The total value of Russia's subsoil assets according to the estimate by the Ministry of Natural Resources amounts to circa $28 trillion.
Taking into account the world market prices for 1999-2000, the share of different kinds of mineral resources in the aggregate value of Russia's subsoil assets is as follows:

- Fuel and energy resources - 72.4%.
- Non-metallic mineral resources (including agro-ores) - 13.6%.
- Mineral resources for ferrous metallurgy - 6.6%.
- Mineral resources for non-ferrous metallurgy - 6.1%.
- Rare and trace rare metals - 0.7%.
- Noble metals and diamonds -1.1%.

The distribution of the mineral resources throughout the Russian territory is extremely non-uniform (see figure 1). The most significant natural resources are concentrated in the Uralian federal okrug, which includes oil- and gas-bearing basins of the Western Siberia, Yamal-Nenets okrug, and Uralian mining regions with deposits of Fe, Cu, Au, bauxites, rare metals.

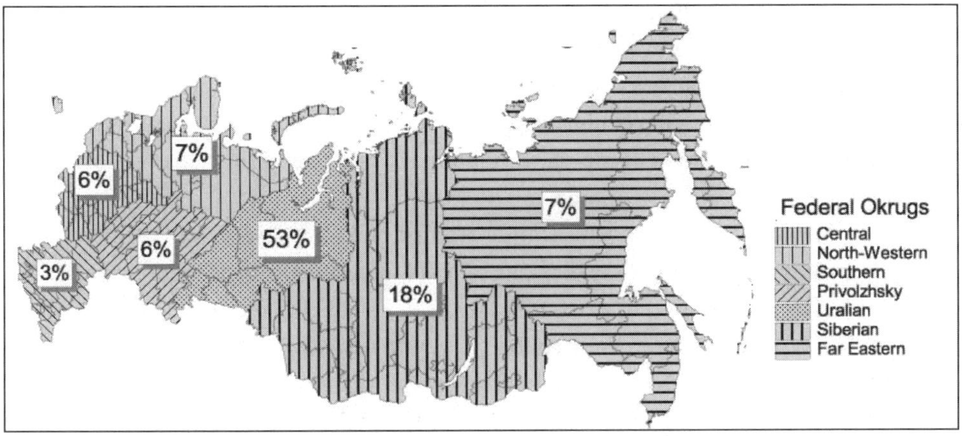

Figure 1. Distribution of estimated value of mineral resources by federal okrugs

Judging by the results of the latest prospecting operations, the importance of the North-western federal okrug is to increase dramatically in the nearest future. In this very region there are exposed the prospects of considerably expanding the mineral resource base of the varieties of chief raw materials, which are in the lowest supply in Russia – viz. titanium (Pudozhgorskoye deposit, Karelia), chromium (Moncha, Kola Peninsula; Burakowka, Karelia), manganese (Porozhskoye, Komi Republic), platinoids (Fedorovskiye Panskiye Tundry, Kola Peninsula; Burakowka, Karelia), aluminium (bauxites of Timan).

On the contrary, the importance of the Privolzhsky and Southern federal okrugs will be apparently diminishing in the future, in connection with considerable exhaustion (up to 70-80%) of the reserves of hydrocarbon mineral materials.

There are great prospects for the successful development of the Siberian federal okrug at the expense of developing large oil and gas fields of Evenkia (Krasnoyarsk Territory), gold-ore occurrences of the Yenisei Ridge, and of further expansion of intense prospecting operations on the Taimyr Peninsula.

The main problems of the Far Eastern okrug consist in the necessity to restore the practically blighted mines and integrated plants of tin, wolfram, boron, antimony, beryllium, and fluorite deposits of Sikhote Alin, Priamurye, Yakutia, and Chukotka and in that of further development of oil- and gas-bearing basins of Sakhalin shelf, where the Sakhalin-1,2,3 fields are already operating.

One of the characteristic and, at the same time, extremely favourable for the future development, features of the mineral resource base of Russia is the high percentage of unique — superlarge and large by their reserves — deposits. Such occurrences have rather high value, they allow planning their long-term efficient development with due regard to all contemporary environmental and social demands. The total number of such largest deposits, defining the country's present mineral-resource visage, amounts to circa 100-120 oil and gas fields and the same number of those of chief varieties of solid mineral resources.

Notable with regard to deposit value rating are first and foremost oil and gas deposits, and the absolute maximum of the deposits of solid mineral resources, with the Norilskoye Ni, Cu, Co, Pt, Pd, Au, Se, Te, Ag, S deposit being the second largest in the world (after Witwatersrand in the Republic of South Africa). To the most valuable deposits belong also: those of potassic salts (Solikamsk), apatite (Khibiny, Murmansk region), iron ores (KMA, central regions), tin (Deputatskoye, Yakutia), gold (Olimpiadninskoye, Krasnoyarsk Territory), uranium (Streltsovskoye deposit, Chita region – the third place in the world), diamonds (kimberlite pipes of Yakutia and Arkhangelsk region, the latter ones are not yet being developed; as well as those of Popigai shock crater in Dolgan-Nenets Autonomous Region), and others.

To the occurrences of very high value belong also yet undeveloped deposits of gold - Nezhdaninskoye (Yakutia), of gold and platinum - Sukhoi Log (Irkutsk region), of copper - Udokana (Chita region), of boron - Dalnegorskoye (Primorski Krai), of lead and zinc - Kholodninskoye (Irkutsk region) and Gorevskoye (Krasnoyarsk Territory), of rare metals and rare earths, including the largest ones in the world – Tomtor with very high-grade ores (Tr, Nb, Sc) in Yakutia, Katuginskoye (Ta, Tr, Nb, Zr) in the Chita region, and a number of others.

If one should regard more specifically the state of things by separate groups of mineral resources, so in such case the Russian economy is, naturally, determined by abundant fuel and energy resources — the largest reserves of natural gas, petroleum and coal. Export of gas and oil, mainly to the Western Europe, produces circa 50-60% of the country's foreign exchange earnings.

What are the prospects of further development of Russia's oil and gas complex?

The main oil and gas resources, which are being actively developed today, are concentrated in the Western Siberia, the regions adjacent to the western shoots of Ural Mountains (Priuralye), including Tymano-Pechorskaya lowland, the northern coastal regions of the Caspian See (Prikaspiye), and in the Northern Caucasus (see figure 2).

Figure 2. Largest Russian mineral deposits

One of the most important energy resources of the 21st century will be the shelf of northern seas, where there are already discovered, but yet not developed, the vast fields of gas condensate, viz. the Shtokmanovskoye field in the Barents Sea, Rusanovskoye and Leningradskoye in the Kara Sea, Varandei-more on the Pechorski shelf, and others. Furthermore, there is beginning the intense development of the unique oil and gas deposits of the Eastern Siberia, Evenkia (Yurubcheno-Tokhomskaya group), large deposits of Yakutia (Chayandinskoye, Sredne-Botoubinskoye), of the Sea of Okhotsk (the shelf of Sakhalin).

In the Western Siberia the main reserve for the future is the Yamal Peninsula with the unique oil deposits (Russkoye, Komsomolskoye).

Analysis of the oil- and gas-bearing capacity prospects witnesses Russia's considerable potential resources.

At the same time, under actual present circumstances the situation as the result of the exhaustion of the developed deposits of Priuralye and the Western Siberia is rather alarming and may lead to a decrease of oil production within the next five years.

Another problem facing the Russian economy is abnormal balance of the most important energy resources — rather unduly high proportion of gas in the power industry (more than 50%, generally in the world circa 24%) and unduly low onc of coal and combustible shales (~ 11-12%, in the world – 26.2%).

In the past decades coal production in Russia sharply decreased at the expense of the utilization of cheap gas. In prospect, however, changes are certain to take place. It is necessary to take into consideration large reserves of the high-grade coals of Kuzbass and of the Kansko-Achinski (Eastern Siberia) basin, as well as large Vorkuta and Yakutian coal basins, which under the new exploitation technologies (simultaneous extraction of methane and admixed elements, the using of water and coal mixtures for the transportation of coals, and others) may efficiently be used and make up certain proportion of the export of energy resources. By their ranks and grades the coals of the

Eastern Donbass are not inferior to the coals of the Silesian basin of Poland and are comparable to the actively exploited coal basins of the USA (Appalachians and Illinois).

4. The problems facing development of mineral resource base

All the abundance of the mineral resources notwithstanding, we must, however, take into consideration the fact, that in Russia there are objectively the most unfavourable natural conditions for their exploitation, compared with other countries — more than $^2/_3$ of all mineral deposits are located within the zone of ever-frost, considerable territories, including oil- and gas-bearing ones, as well as the largest fields of copper-nickel, apatite and other most important kinds of mineral resources are situated beyond the Polar circle. Alaska is the only place in the world, where exploitation is being carried out under similar conditions with respect to the unfavourability.

Notwithstanding considerable reserves of mineral resources and well-developed mining and process industries, the share of Russia in the extraction of mineral resources has sharply decreased in the last ten years, which fact is especially significant against the background of substantial growth of mineral resource production in the world (table I). The only exception is the production of diamonds (20% growth – at the expense of Yakutian deposits), and that of platinoids (at the expense of Norilsk deposits). There are different causes of the decrease in production: firstly, this is the decay of traditional economic ties, which existed in the USSR, secondly, with the transition to the market economy, the requirements were sharply toughened in regard to the quality of mineral raw materials, contents of mineral resources, accessibility for exploitation, efficiency of the technologies used, and other characteristics, determining the manufacturing cost of the production. To a great extent the production decrease is also a result of the paradoxical situation, which has formed in the recent years, when as a consequence of ill-conceived taxation policy, and disproportionate increase of prices during 1990-1999 (price of electric power was increased 24 thousand times, that of freight traffic – 20 thousand times, and of petroleum products – 16 thousand times), it became unprofitable to exploit many deposits of northern regions of Siberia and the Far East. Even the extraction of gold from the primary deposits of Yakutia without updating production lines is unable to sustain competition, as the manufacturing cost becomes higher than the price of gold on the London Stock Exchange. Eventually in Russia it became non-remunerative to exploit many primary gold-ore deposits. In a number of cases it also proved to be economically unprofitable to carry on mine-shaft extraction of coal and uranium by means of underground workings, and so forth. As a matter of fact, it was because of the exploitation's being unprofitable, that many mining plants with expensive equipment and well-developed infrastructures were closed down, including such well-known ones, as Pevekski (Sn) and Iultinski (W) Metal Enrichment Plants in the north-east, Sherlovogorski (Be, W) and Orlovski (Ta, Nb) MEPs of Zabaikalye, Tuvakobalt integrated plant, Deputatski (Sn) MEP in Yakutia, and many others.

TABLE I. The production of mineral resources in Russia and in the world

Kinds of mineral resources	Russia	The World		
	Increase	Decrease	Increase	Decrease
Diamonds	20	-	25	-
Gold	-	22	13	-
Silver	-	50	8	-
Petroleum	-	33	11	-
Gas	-	9.5	13	-
Coal	-	3	-	-
Iron ore	-	11	10	-
Bauxites	-	15	8	-
Copper	-	20	30	-
Nickel	-	6	-	5
Lead	-	77	-	1
Zinc	-	30	2	-
Tin	-	73	8	-
Wolfram	-	75	-	40
Molybdenum	-	70	24	-
Antimony	-	90	28	-
Titanium	-	40	29	-
Niobium	-	80	33	-
Tantalum	-	95	13	-
Apatites	-	50	-	-
Phosphorites	-	83	-	-
Potassic salts	-	20	-	-

Thirdly, it must be borne in mind that the present state of the mineral resource base of Russia is fundamentally different from what we had in the Soviet Union. By the year 2000 (the beginning of the century) practically all previously discovered and attractive by their economic parameters mineral deposits, mines, and metal enrichment plants were privatized. At the same time, the greater part of deposits was transferred in 1992-1993 to the utilizers of mine resources free of charge (circa 64% of oil reserves, 63% of gas reserves, 86% of reserves of diamonds, 71% of reserves of gold and other mineral resources to the amount of hundreds of billions of US dollars). According to the data of the Ministry of Natural Resources, today circa 90% of all industrially valuable installations are licensed. Out of total number of licences issued for the exploitation of mineral resources, the maximum quantity falls on gold, platinoids, diamonds, oil, and gas (90% of reserves of oil, 80% of gas, 86% of diamonds, 80% of gold are licensed). In the majority of licenses, granted free of charge during 1992-1993, there were not specified the conditions for the utilization of subsoil assets. At the same time, the general situation of unsteady economic development objectively encourages, on the one hand, robber-exploitation of the deposits – the choice of the most high-grade ores, the utilization only of easily-extracted portions of mineral reserves, and on the other hand – the «freezing» of the exploitation of deposits «until the better days», or for the «resale» of licence. At the same time, the substantial difference between the prices on the internal and foreign markets quite objectively causes the mineral resource utilizers' being interested in exports. The characteristic examples are: the price of oil on the internal market (for the year 2000) is 29-32% of the world price; of gas – 15-17%, of potash fertilizers – 17-18%. It is only too natural, that the utilizers of mineral resources are seeking to penetrate the foreign market and to raise prices on the internal one, which is practically impossible seeing the standard of wages existing in Russia.

Illustrative in this regard are the data, attesting the fact that the country, having the richest deposits of agrochemical raw materials – the unica of world-wide importance like Khibiny (apatites) and Verkhnekamskoye deposit (potassic salts – the first place in the world by reserves), occupies one of the last places in Europe by fertilizer usage in agriculture. Properly speaking, the demand for fertilizers in the country is met only by 10%! As a result of this, the country which is so rich in natural resources stands on a very low level of the consumption of its own raw material resources.

In connection with the transition to the market economy, in the development and exploitation of mineral resources there can be distinctly observed a number of complicated and multifaceted problems, upon the solution of which the evaluation of the prospects of the development of mineral resource potential of Russia and the country's stand in the XXI century world economy wholly depends.

The main problem consists in the absence of an adequate reproduction of the reserves of mineral resources owing to the sharp curtailment of the funds, allotted from the state budget for regional geological surveying operations and for thematic and prospecting investigations.

In this connection the exhaustion of the active reserves which are in the balance now will start already within the next decades.

The prospected reserves of uranium, manganese, chromium, barium, kaolin, and bentonite are very limited. To discover new fields and to ensure a considerable accretion of reserves is the task for the quinquennium immediately ahead. The situation remains also tense in respect of the reserves of bauxites, titanium ores, molybdenum, tin, flogolite, and ore (primary) gold.

Easily discernible are also the problems in maintaining the existing level of the extraction of oil, copper, lead, zinc, antimony, niobium, germanium, lithium, rare earths. The operating deposits are considerably exhausted, and the development of new ones requires major investments and solution of many social and environmental problems.

Satisfactory enough in respect to the supply of reserves for more than 10-15 years is the situation with the following mineral resources: gas, coal, diamonds, iron, nickel, platinum, beryllium, potassic salts, and asbestos.

Considering the importance of Russia's natural resources in the long view, it is necessary to be aware of the major world-wide trends, which Russia has to take into account.

First and foremost we must lay emphasis on the permanently increasing demand of developed and developing countries for mineral resources. Nowadays the USA – 4% of the global population – consumes 20% of mineral resources, at the same time they also exploit their own mineral resources 2-3 times more effectively than Russia.

16% of the population of «the Golden Billion» consumes more than 52% of the world's reserves of mineral resources.

In prospect, although the consumption growth will diminish in respect of rate, it will be increasing steadily, first of all as a function of population growth. In the estimation of sociologists, the increase of population up to 7.5-10 billion men by the year 2050 will bring about the necessity of improving the present standard of consumption of mineral resources by 25% (The population has increased fifteen times in the last 500 years and amounts now to 6.1 billion men).

Academician I.S. Gramberg in his work, dedicated to the problems of development of fuel and energy resources, published in 1998 in the journal «Geology of Oil and Gas», adduced two scenarios of prognoses, made by the World Energy Council (WEC). According to one of them, in 2020 world consumption of power resources will increase twofold (from 12.5 to 24.7 billion tons of fuel equivalent), according to another, the minimum one – from 12.5 to 16 billion tons of fuel equivalent. And this is with USA, Canada, Great Britain, and Norway being provided with their own extractable reserves of oil only for 6-8 years, and of gas – for 14 years. It becomes clear, why academician N.A. Shilo concluded: "A keen struggle for the planet's resources awaits us".

Another main world trend in the use of natural resources consists in an ever-increasing integration of natural resource potentials of most countries of the world, globalization of economy with an increasingly important role played both by world-wide (International Monetary Fund, World Trade Organization) and regional (European Union, Association of South East Asian Nations) or special (OPEC and others) organizations. A distinctive feature of the globalization is the consolidation of companies and the establishment of transnational holdings, operating on the installations of the whole world. This circumstance provides for such companies the possibility to have "much room for manoeuvre" for efficiently distributing funds by installations and influencing the level of world prices on resources. With such schedule the national interests are regulated solely by the State and its laws.

None the less distinct is the tendency towards permanently growing environmental and social pressure on the value of extracted mineral resources, which makes it profitable to exploit only the largest deposits of the world with explored reserves for ten and more years. Orientation to large, superlarge, and giant deposits found its expression in the concept of sustainable development (Rio de Janeiro, 1992), and the topics of the proceedings of the International Union of Geologists and the International Programme for Geological Correlation. It should also be stated, however, that for the next 5-10 years it would be just and economically profitable for Russia to additionally develop small business on small-scale and medium, but high-grade deposits with the use of brigade method, tour method, and prospector method of operations.

Finally, it is necessary to take into consideration the determinate and consistent reduction of world prices on mineral resources (Sn, U, Au, and others), which is connected with the improvement of technologies of the extraction and processing of mineral resources. This last circumstance allows the profitability only of those plants, which are actively introducing new knowledge-intensive technologies.

An exception to the common picture of price reduction may be palladium and rhenium out of metals and the extraordinary 2-3 times jump in oil prices, which took place in 1996 and allowed us to improve dramatically the economic situation in the country.

All these factors, determining the development of the mineral resource base of the countries of the world and their integration into the world-wide economic system, first of all depend (value of shares, volume of investment, etc.) upon the stability of economic policy of states and transparency of the economic activities of companies along with providing full account of the most significant indices through the Internet.

These very factors determine nowadays successes of some countries – like USA, Australia, Canada, or Norway, and the "troubles" of others – the developing countries, rich in natural resources, and this is what first and foremost limits the development of Russia's mineral resources.

Inevitability and justifiability of the struggle for the world resources are so substantial, that they should determine the whole strategy of Russia's economic development, first of all – harmonious development of natural resources and high technologies in the mining industry.

5. Conclusion

Eventually, all prospects of long-term development of mineral resources of Russia and its role in the world economy depend today upon the resolution of the above-mentioned and many other difficulties, which arose at the turn of the century after the collapse of the Soviet Union.

To resolve them, a fundamental improvement of legislation is necessary. The enacted "Laws on Subsoil Assets" have only drawn the outlines of the solution of problems and require numerous alterations, amendments, and the sharp increase of the role of the State regulation in the field of study, development and exploitation of mineral resources.

MINERAL DEPOSITS OF TURKEY

T. ENGIN
MTA Genel Müdürlüğü, Maden Etüt ve Arama Dairesi, Ankara, Turkey.
E-mail: tandoganeyahoo.com

1. Introduction

Archaeological studies show that history of mining in Turkey goes back to 7000 BC. Mining appears to have played an important part in the development of the ancient civilizations in Anatolia, such as those of Phoenicians, Hittites, Phrygians, and Lydians. Some mines operating even now in Turkey were originally discovered and mined by the people of the ancient civilizations.

Turkey is located within the alpine orogenic zone and has been subjected to intense tectonic movements. Turkey has been divided into numbers of tectonic plates, intensive igneous activities developed and ophiolite emplacements took place. All these activities made Turkey's geology complicated. This complicated geology is the cause of rich variety of Turkey's mineral wealth. Copper, lead, zinc, iron, chrome, mercury, sepiolite, borate, are the traditional minerals of Turkey. Data related to Turkey's minerals are given in Tables IV-XIII (See in Appendix).

There are 90 minerals in the world, which are traded in the market. Out of these 90, 73 minerals are mined in Turkey. 17 of these 90 minerals are not present in the country. Reserve and production level of 23 items are not at the required level. Turkey has source superiority for 50 minerals and ores.

For the preparation of metallogenic map of Turkey data on 3384 copper, lead zinc, iron, manganese, chrome, nickel, wolfram, gold, silver, aluminium, antimony, mercury, pyrite, barite, magnesite, kaolin, asbestos, graphite, trona, thorium, talc, phosphate, borate, sulphur deposits have been gathered. To prevent unnecessary crowding, mineralizations close to one another and having common characteristics have been grouped and 1672 deposits or mineralizations have been plotted on the map.

Metallogenic map of Turkey shows that there are numbers of structure and rock type controlled metallogenic provinces. Numbers of metallogenic provinces with different age and origin are noticeable (see figure 1).

A.G. Tvalchrelidze and G. Morizot (eds.),
Mineral Resource Base of the Southern Caucasus and Systems for its Management in the XXI Century, 81–104.

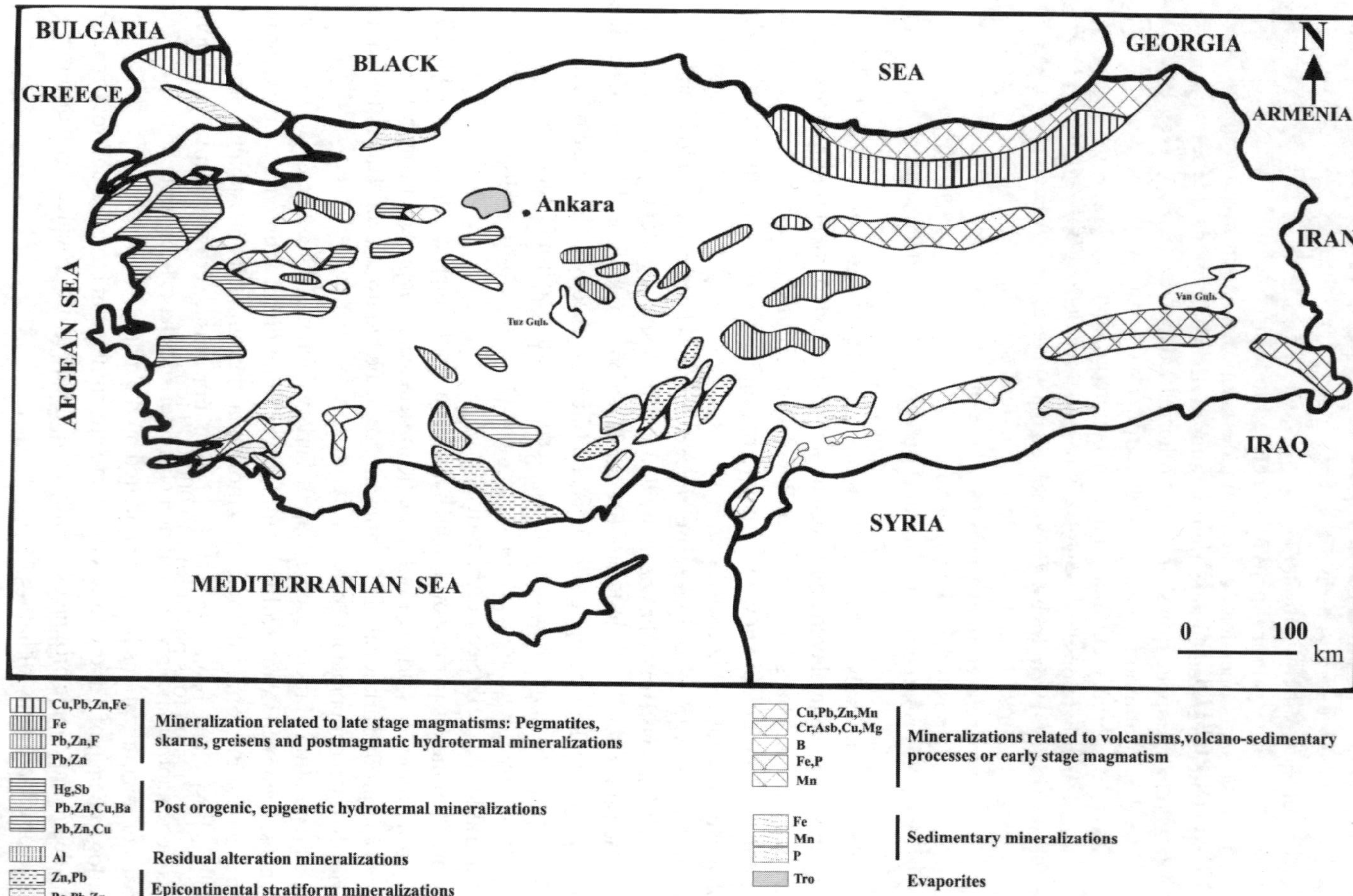

Figure 1. Some metallogenic zones in Turkey

Some of the metallogenic provinces in Turkey are:
Mineral provinces related to ophiolites. They include chromite, magnesite, asbestos, copper deposits and mineralizations.
Volcano-sedimentary mineral provinces related to felsic volcanics. They include copper, copper-lead, copper-lead-zinc, manganese deposits and mineralizations.
Mineral provinces related to acid-neutral magmatisms. They include copper, copper-lead-iron, porphyry copper-iron skarns-hydrothermal copper-lead-zinc, copper-lead, wolfram, lead-zinc, antimony-gold, antimony, iron-wolfram-molybdenum, iron, fluorite, copper-zinc-lead, mercury-antimony, antimony-mercury, mercury, barite, iron-phosphate deposits and mineralizations.
Sedimentary mineral provinces. They include manganese, iron, red bed type iron-copper, phosphate, strata-bound barite-lead, zinc-lead deposits and mineralizations.
Mineral provinces related to evaporates. These provinces include borate, trona, salt, gypsum, celestite deposits and mineralizations.
Mineral provinces related to laterites and placers. They include bauxitic iron, bauxite, nickel, gold deposits and mineralizations.

2. Metallic Minerals in Turkey

2.1. IRON ORE DEPOSITS

Iron ore deposits in Turkey are mainly small and medium-sized and are genetically of 6 types, viz.

- Contact metasomatic iron ore deposits: Divriđi A, Divriđi B (Sývas, C. Turkey)7; Yahyalý-Karamadazý (Erzincan, E.Turkey); Býzmýţen,Hasan Çelebi (Malatya, E. Turkey).
- Hydrothermal metasomatic iron ore deposits: Attepe, Mansurlu (Adana, S. Turkey).
- Volcano-sedimentary iron ore deposits: Hekimhan (Malatya, E. Turkey), Büyük Eymir (Balýkesir, W. Turkey).
- Marine-sedimentary iron ore deposits: Çamdađ (Sakarya, NW. Turkey).
- Lateritic iron ore deposits: Avţarören (Kangal, Sývas C. Turkey), Karaçam (Eskiţehir, W. Turkey).
- Placer iron ore deposits: C placer (Divriđi, Sývas, C. Turkey), Eastern Black Sea coastal sands.

Important iron ore deposits are mainly of contact metasomatic type. They are mainly located in Central Turkey around Sývas, Kayseri, Malatya regions.
Minable ore reserves with 55% average Fe grade are estimated to be (101 500 000 tons proven + 49 000 000 tons possible) 150 000 000 tons.

2.2. COPPER-LEAD-ZINC DEPOSITS

Copper-lead-zinc deposits in Turkey are genetically of 5 types. They are:

- Deposits related to ophiolites.

- Deposits related to andesitic-dacitic volcanisms.
- Skarn and vein type deposits.
- Porphyry type deposits.
- Strata-bound deposits.

Economically important copper deposits in Turkey are the ones related to basic section of ophiolites also known as Cyprus type and also the ones related to andesitic-dacitic volcanics known as Kuroko type.

Ophiolite related copper deposits are mainly located in the SE. ophiolite belt and in Küre region, N. Turkey. Kuroko type deposits are mainly located in the Northern Black Sea region

Skarn mineralizations related to granitic and dacitic rocks are important from the point of view of lead-zinc mineralizations. They are mainly located in NW. Turkey in Balýkesir, Çanakkale areas, in Yozgat (C. Turkey), Ţebinkarahisar Giresun, N. Turkey) areas.

There have been several porphyry copper mineralizations discovered in Turkey. They are related to granitic-granodioritic intrusions. Some of those mineralizations are:

TABLE I.

Location	Reserve (in tons)	Grades (%)	
		Cu	Mo
Bakýrçay (Merzifon, N. Turkey)	200 000 000	0.2	
Ulutaţ (Ýspir, Erzurum, E. Turkey)	140 000 000	0.27	0.18
Dereköy (Kýrklareli, NW. Turkey)	220 000 000	0.25	0.003
Ýkiztepe (Kýrklareli, NW. Turkey)	13 000 000	0.39	0.005
Güzelyayla (Maçka, Trabzon, NE. Turkey)	115 000 000	0.18	0.012
Topukdere (Domaniç, Bursa, NW. Turkey)	120 000 000	0.17	

None of these mineralizations have been found to be economically exploitable.

Strata-bound carbonate type lead-zinc deposits are located at certain levels of Palaeozoic-Lower Mesozoic limestones in the Taurus Mountains along S. Turkey. In respect to reserve they are small, but are of high grade.

Estimated copper reserve of Turkey amounts to 93 300 000 tons (proven + probable) at 2.47% Cu grade. It is equivalent to 2 280 000 tons of metal copper. When reserves of lower grade copper deposits are included, metal copper reserves of Turkey go up to 3.5 million tons.

Sulphur type lead-zinc reserves of Turkey are estimated to be 68 700 000 tons (37 500 000 tons proven + 10 500 000 tons probable + 20 500 000 tons possible) at 5.18% Zn, 1.66% Pb, 3% Cu grades.

Oxide type lead-zinc reserves are estimated to be 860 000 tons (330 000 tons proven +

260 000 tons probable + 270 000 tons possible) at 17.3% Zn, 6.7% Pb grades.
In other words, total metal zinc and lead reserves of Turkey amount to 2 300 000 tons and 860 000 tons respectively.

2.3. CHROMITE DEPOSITS

Chromite deposits in Turkey are of so-called alpine type. They are comparatively smaller in size than stratiform types but are of higher grades. They are located in the tectonite section as well as in the ultrabasic sections of the cumulates of the ophiolite assemblages.
Geographically they are located in 6 major areas. In respect to potential, in relative order of importance, they are:

- Guleman region (Elazýđ, E. Turkey).
- Kopdađ region (Erzincan-Erzurum, E. Turkey).
- Fethiye-Köyceđiz-Denizli region (SW. Turkey).
- Bursa-Kütahya-Eskiţehir region (NW. Turkey).
- Mersin-Aladađ-Pýnarbaţý region (S. Turkey).
- Ýskenderun-Kahramanmaraţ region (S. Turkey).

Chromite deposits are made of chromitite bodies of various size. Their relation with the host peridotite is complicated. Exploration along with mining is the best way of finding new bodies. Because of this reserve estimation studies of alpine type chromite deposits carry some degree of ambiguity.
Chromite mining in Turkey started in 1850. Since then Turkey has an important part in the world-wide chromite market. Since the beginning of chromite mining in the country, Turkey's production is estimated to be about 45 million tons.
Turkey's production is mainly export-oriented. Depending upon price variations, chromite production goes up or down accordingly. The production figures of the last 5 years are given in Table II.

TABLE II.

1995	2 080 043 tons
1996	1 279 032 tons
1997	1 646 413 tons
1998	1 440 470 tons
1999	1 019 542 tons

Turkey has 2 ferro-chrome plants and 1 chrome chemicals plant operating. Their annual chrome ore need is about 530 000 tons.
Taking the available data into consideration, chrome ore reserve of Turkey is estimated to be about 26 million tons. Batý Kef chromite deposit (Guleman, E. Turkey) with 6500000 tons reserve at 20-35%Cr_2O_3 grade is the largest deposit in the country. On the other hand Aladađ (Adana, S. Turkey) low-grade chromite deposit with 200 000 000

tons reserve at 5.42% Cr_2O_3 grade is a source, waiting for being exploited. Technological tests carried out on this deposit indicated that the ore could be upgraded to 50% Cr_2O_3 with 70% recovery.

2.4. NICKEL DEPOSITS

Although ultrabasic and basic rocks cover large areas in Turkey and extensive lateritic bauxite deposits are located along the Taurus Mountains of S. Turkey, still there is not any operating nickel deposit in Turkey. The presence of lateritic bauxite deposits shows that in the geological past, purely climatologically, Turkey was subjected to tropical-subtropical climates. It appears that these climatic conditions caused the enrichment of aluminium but not of nickel.

Çaldađ (Manisa, W. Turkey) nickel deposit with 49 million tons reserve at 1.37% Ni grade is the only lateritic nickel deposit in the country.

Pancarlý nickel sulphide occurrence with 4% Nickel, 2%Cu grades in the Bitlis metamorphics in E. Turkey is yet to be explored.

2.5. ALUMINIUM DEPOSITS

Aluminium deposits in Turkey are mineralogically of 3 types and are concentrated in 10 different areas. They are: Seydițehir-Akseki, Silifke-Tațucu (S. Turkey), Zonguldak (N. Turkey) deposits, which are böhmitic; Ýslahiye-Payas, Yalvaç-Țarkikaraađaç deposits, which are iron bauxite and Milas-Muđla, Alanya, Ýçel-Bolkardađ, Tufanbeyli-Saimbeyli (S. Turkey) deposits, which are diasporitic bauxite ones.

Diasporitic bauxites from Milas area are exported for grinding purposes.

Deposits in Southern Turkey supply material to the aluminium plant in Seydițehir (S. Turkey).

Iron bauxite deposits in Yalvaç and Payas have technological problems.

Aluminium reserve of Turkey is estimated to amount to 85 million tons at 55% Al_2O_3 grade (25 million tons of metal).

2.6. GOLD AND SILVER DEPOSITS

Although the river Pactole was the place from where Lydian king Croesus made his fortune from the gold content of the river, yet at present there is no operating gold mine in Turkey.

With the increase of the price of gold in the world market in the 70s and the development of the heap leaching technology there was a surge of gold exploration in Turkey towards the end of 1980s.

Turkey's geology is quite favourable for epithermal gold mineralizations. Tertiary acidic and neutral volcanics cover large areas in the country. These volcanics have been subjected to various degrees of alterations in many places.

A new mining law, giving foreign companies an incentive to form joint companies with

the Turkish counterparts in order to go into mining in Turkey came into effect in 1985. As a result, foreign companies as well as relevant public organizations, such as MTA and Etibank, run extensive gold exploration programmes.
These exploration activities soon proved fruitful. At the moment 9 economically viable gold deposits have been discovered (see in table III).

TABLE III.

Deposit	Grade(gr/t)		Reserve(ton)	Metal content (ton)	
	Au	Ag		Au	Ag
Ýzmir, Bergama Ovacýk, W. Turkey	9	11	2 980 000	26.82	32.78
Ýzmir,Seferihisar,Efemçukuru,W.Turkey	12.65	-	2 500 000	31.62	-
Uţak, Eţme, Kýţladađ, W. Turkey	1.43	-	74 000 000	105.8	-
Balýkesir,Havran,Küçükdere, W. Turkey	6.43	11.8	1 410 000	9.07	16.64
Eskiţehir,Sivrihisar,Kaymaz, C. Turkey	6.04	5.3	974 000	5.88	5.17
Gümüţhane,Mescitli,Mastra, E. Turkey	12	-	1 000 000	12	-
Çanakkale,Kirazlý,Akbaba, NW. Turkey	1.25	-	8 000 000	10	-
Artvin,Cerattepe,NW. Turkey (iron cap)	4	140	8 200 000	32.8	1148
Massive sulphide	1.2	25	3 900 000	4.68	97.5
Total				**240**	**1300**

None of these deposits have been mined so far because of the environmental concerns, which have brought gold exploration as well as some other mining activities in the country to a standstill.
It is believed, that with the start of gold mining in the country more gold deposits will be discovered and mining in general will get a fresh boost in Turkey.
Gümüţköy (Kütahya, W. Turkey) silver deposit is the only operating silver mine in Turkey. It is a hot spring type epithermal mineralization.
The estimated reserve is about 21 500 000 tons at 180gr/ton and 0.2gr/ton silver and gold grades respectively. This is equal to 3800 tons of metal silver and 4 tons of metal gold.

2.7. MANGANESE DEPOSITS

There are numerous manganese occurrences in the country. Except for the sedimentary manganese deposits of Binkýlýç (Çatalca, Ýstanbul, NW. Turkey) and Ulukent (Denizli, SW. Turkey), all other deposits are of volcano-sedimentary type.
Ulukent (Denizli) deposit with the reserve of 2 500 000 tons at 31% Mn, 5% Fe, 20% SiO_2 is the biggest in the country. Mineralization is located along the contact between underlying Upper Jurassic chert bearing limestone and black limestone. Main ore

mineral is rhodochrosite. All other manganese deposits, which are small, with the reserves of each of them not exceeding several thousand tons, are of oxide type. Ulukent is the only deposit of carbonite type.
Manganese reserve of Turkey is calculated to amount to 4 500 000 tons at 34.5% Mn grade (1 600 000 tons of metal).

2.8. WOLFRAM DEPOSITS

Wolfram occurrences are located in Uludađ, Kozbudaklar (Bursa, NW. Turkey), Gümüţler (Niđde, C. Turkey) and Kebandere, Sođanlýköy, Nallýziyaret (Keban, Elazýđ, E. Turkey). They are situated in the skarn zones along the contact between intrusives and crystalline limestones.
Among those only Uludađ occurrence is at deposit scale. The reserve of Uludađ deposit is calculated to be equal to about 37 000 tons of metal. The deposit has technological problems. In the past it was mined for some time but because of the technological difficulties and unattractive market price it was abandoned in 1990.

3. Non-metallic Minerals in Turkey

A large part of Turkey was subjected to the shallow water sedimentation of the continental type under the regional climatic conditions in Oligocene. There developed widespread evaporatic borate, natural soda, sodium sulphate, celestite, rock salt deposits in Miocene and Pliocene.
Turkey is in some sort fortunate in respect of varieties and reserves of industrial minerals. At present mining activity is being carried out on more than 30 industrial minerals. Borate deposits are located in NW. Turkey (Balýkesir, Kütahya, Bursa, Eskiţehir). The source for Boron is thought to be calc-alkaline volcanisms. Borate mineralization is present in the clay, marl, tuff association. Miocene limestones underlie, Palaeocene limestones overlie the borate zone. Main borate minerals are colemenite, oulexite, hydroboracite. Ore reserves explored so far at 26-35% B_2O_3 grade amount to 2 200 000 000 tons, which is 60% of the world's reserve.

3.1. TRONA AND SODIUM SULPHATE DEPOSITS

Trona and Sodium sulphate deposits occur in the Beypazarý-Çayýrhan Neogene basin, NW. Ankara. Upper Cretaceous-Upper Pliocene volcanisms are believed to have supplied material to the Neogene lakes. Trona deposits occur in the bituminous shale, claystone, tuffite, dolomitic marl of Middle-Upper Miocene. The thickness of pure trona horizons varies from 2-3 cm to 11.5 m within the 40m thick ore zone. Besides trona pirssonite, nahcolite, gaylussite may occasionally be found in the ore zones.
Total reserve of the deposit is calculated to amount to 235 million tons with 56% trona. This deposit is the second biggest deposit in the world after the Wyoming deposit of the USA.

3.2. SODIUM SULPHATE

Sodium sulphate deposit near Çayýrhan was also formed in the same Neogene basin as trona. Gypsum deposits are interlaid with siltstones, claystones, mudstones. The thickness of the gypsum horizons varies from 1.85 m to 20 m. The reserve is estimated to amount to 3.5 million tons with 63-89% Na_2SO_4 and 192 million tons with 30% Na_2SO_4 contents.

3.3. CELESTITE DEPOSITS

Celestite deposits are located in the Late Eocene, Oligocene, Early Miocene formations of the Sývas (C. Turkey) sedimentary basin. It was formed in the tidal flat and sabkha facies. Two types of mineralizations may be identified:
(a) - Syngenetic, massive and layered deposits;
(b) - Epigenetic replacements in the primary deposits.
Celestite mostly occurs in gypsum and occasionally in limestone, claystone, siltstone. There are about 40 deposits in the Sývas region. Estimated reserve is about 3 million tons.

3.4. ROCK SALT DEPOSITS

Rock salt deposits are located in Central Anatolia; Çankýrý, Yozgat, Nevţehir regions and in Erzurum, Kars regions in E. Turkey. They are timed to Oligocene, Miocene, Pliocene. Total rock salt reserve amounts to 5.7 billion tons.

3.5. CLAY DEPOSITS

3.5.1. *Kaolin deposits*

There are about 50 kaolin deposits in Turkey. They are located in Northwest, Central and Black Sea regions of Turkey. Northwest and Central Anatolian kaolin deposits are related to the hydrothermal alteration of Neogene andesites, dacites and rhyolitic tuffs. In the Northern Black Sea region they are related to the alteration of Upper Cretaceous dacites.

Turkish kaolins are hard, they generally contain fine grains of quartz and feldspar. Alunite and iron oxide contents are also high.

Düvertepe (Sýndýrgý, Balýkesir) is the biggest deposit in the country with 63 million tons of reserve.

Sedimentary clay deposits are mainly located in the NW. Turkey around Ţile, Ýhsaniye (Ýstanbul); Söđüt (Bilecik); Yenice (Çanakkale). They are located in the Neogene formations overlying Palaeozoic clayey schists, greywackes. Kaolinitic clay deposits occur as several horizons on and under the coal seams.

Clay reserve of Ţile deposit (Ýstanbul) is estimated to amount to 230 million tons, reserve of Ýhsaniye deposit (Ýstanbul) amounts to 25 million tons.

3.5.2. *Bentonite deposits*

Bentonite deposits occur in various parts of the country and are related to Upper Cretaceous and Neogene volcanics. They were formed by diagenetic alteration. Reţadiye (Tokat, North-eastern Central Turkey) Na bentonite deposit has the reserve of 20 million tons. Hançýlý (Kalecik, Ankara) Na bentonite deposit has the reserve of 75 million tons. They are used for drilling, mud and moulding industry and edible oil refining.

3.6. SEPIOLITE DEPOSITS

Sepiolite deposits occur in two forms. One is meerschaum, which was developed by the transformation of magnesite in Neogene basins.

They are mainly used for making ornamental goods. The deposits are located in the Eskiţehir region, W. Turkey.

The second type is the sedimentary sepiolite. They developed as chemical precipice in the alkaline lakes or in the seasonal playa lakes in the arid and semiarid conditions in Upper Miocene-Pliocene. The thickness of the sepiolite lenses may vary from 0.2 m to 1.5 m. In some places it goes up to 7-8 m.

They are mainly used as pet litter, for high technological applications such as cordierite ceramics and paper.

3.7. MAGNESITE DEPOSITS

Two genetic types of magnesite deposits are present in the country:

(a) - Cryptocrystalline/gel magnesite;

(b) - Sedimentary magnesite.

Cryptocrystalline magnesites are related to serpentinized peridotite and are located in Kütahya, Bursa (NW. Turkey) and Konya (C. Turkey) regions. They are of stock work and blumenkohl kinds.

Sedimentary magnesite deposits are located in Erzincan (E. Turkey), Konya (C. Turkey) and Denizli (W. Turkey) regions.

3.8. BARITE DEPOSITS

Barite deposits are located in Gazipaţa (Antalya), Kahramanmaraţ (S. Turkey), in Konya-Isparta (W. Central Turkey) and Muţ (E. Turkey) regions.

Gazipaţa (Antalya) deposits were hydrothermally formed as veins and concentrations in the dolomitic and other carbonate rocks and schists. Estimated reserve amounts to about 5 million tons.

In Muţ area barite occurs as veins in schists, quartzite and marble of the Bitlis massif. The estimated reserve amounts to 3.7 million tons.

In Konya-Isparta region barite occurs in Ordovician-Devonian pelitic and carbonate rocks. Potential reserve amounts to 5 million tons. Total reserve amounts to about 15 million tons, at 62-99% $BaSO_4$ grade.

3.9. FELDSPAR DEPOSITS

Pegmatoids are exploited as feldspar deposits. They are concentrated mainly in Çine, Aydýn (Menderes massif), Simav, Kütahya; Demirci, Manisa (W. Turkey). Albite and orthoclase are the main minerals. In Çine area the estimated reserves of albite amount to 197 million tons at 6-10% Na_2O grade and of orthoclase to 1 million tons at 5-11.5 % K_2O grade.

3.10. PHOSPHATE DEPOSITS

Phosphate deposits are located in Mardin area. Apart from the Mazýdađ (Mardin) deposits there are some other deposits in Bitlis, Bingöl, Gaziantep, Hatay, Adýyaman areas (E. and S. Turkey).
Mazýdađ deposit is in the Cretaceous limestone. Phosphate appears in 4 horizons. Batý Kasrýk horizon is at 19.58% P_2O_5 grade, 38 million tons.

3.11. ASBESTOS DEPOSITS

They are chrysolite and tremolite deposits, which are related to ultrabasic and metamorphic rocks respectively. Asbestos deposits are mainly located in Beypýnarý (Sývas, C. Turkey), Orhaneli (Bursa, NW. Turkey), Kýzýldađ (Hatay, S. Turkey), Mihalýççýk (Eskiţehir, C. Turkey), Çankýrý (C. Turkey), Bitlis (E. Turkey) areas.
They are generally of cross-fibre type and are at the 5th, 6th, and the 7th groups according to the Quebec classification.
Turkey's asbestos reserve is estimated to amount to about 29 million tons.

3.12. MARBLE DEPOSITS

Crystalline carbonate rocks, magmatic rocks, travertines within the metamorphic massifs, Permian and Mesozoic limestones, as well as magmatic rocks such as granite, syenite, serpentinite and diabase are extensively quarried for the production of commercial marble.
Main marble quarries are located in Menderes massif (W. Turkey) and in Konya, Eskiţehir (C. Turkey), Afyon, Bilecik, Bursa, Balýkesir, Çanakkale areas (W. Turkey).
Onyx marble and travertine occur mainly in Turhal (Tokat, N. Turkey), travertine occurs also in Denizli (W.Turkey).

3.13. PUMICE DEPOSITS

Pumice deposits are related to acidic Quaternary volcanisms in Central and Eastern Turkey. Main deposits are located in Nevţehir, Kayseri (C. Turkey), and Bitlis, Ađrý, Kars, Van (E. Turkey) areas.

Thickness of pumice beds vary between 1-25 m. The deposits are overlaid by ignimbirite and basaltic lava flows.
At present pumice production is being carried out by means of the exploitation of the deposits in Kayseri and Nevţehir (C. Turkey) areas.
Pumice reserve is estimated to amount to about 1.5 billion m^3.

3.14. PERLITE DEPOSITS

Perlite deposits are related to Tertiary acidic volcanisms. There are extensive perlite deposits in Ýzmir, Manisa, Balýkesir, Eskiţehir (W. Turkey) and in Bitlis, Van, Kars (E. Turkey) areas. At present production is being carried out only in Ýzmir area.
The reserve is estimated to amount to about 5.7 billion tons.

3.15. QUARTZ SAND, QUARTZ AND QUARTZITE DEPOSITS

Quartz sand deposits of economic significance are found within Eocene and Neogene formations in Tekirdađ and Ýstanbul regions, NW. Turkey. Eocene formation unconformably overlies Istranca metamorphic massif and quartz sand deposits are located at the base of Eocene. They developed as a result of uplift and transgression. The deposits are aligned parallel to the massif and the silica content varies between 75-99%. After some washing and sorting processes, the quartz from these deposits is used in the glass industry.
Neogene quartz sand deposits lie in the area between Kilyos and Terkos to the West of Ýstanbul and Ţile to the East (NW. Turkey). Quartz sands are located at the upper levels of Neogene and show a gradation into clay, coarse sand and silt horizons.
After some washing the sands from these deposits are used in moulding industry.
Quartzite deposits are found in the Silurian-Devonian formations in Adana, Mersin regions (S. Turkey), in Upper Cretaceous formations in Kastamonu, Zonguldak regions (N. Turkey) and in Palaeozoic in Afyon area (W. Turkey).
Vein type quartz occurrences are common in the Menderes metamorphic massif (W. Turkey) and they are used in the ceramics industry.

4. Expected development

At present Turkey's mineral production is within the range of $ 5 billion. Among the miners it is commonly believed, that if mining had some degree of priority in the government's plans, mineral production could easily increase twofold.
Considering the deposits, which have already been discovered and the size, kinds and extensions of the metallogenic belts in the country, one is led to believe that there is still enough room for further discoveries.
It is well understood that the deposits, which have been discovered so far, mostly had

some degree of clear surface manifestations. The deposits yet to be discovered would need more detailed studies and it would cost more.

Turkey has discovered deposits of economic gold, but at present there is no gold mining in the country. With the start of gold mining, gold exploration will get a new impetus. As Turkey has favourable geology for gold mineralizations, it is believed that new exploration work will lead to new discoveries.

The start of gold mining in Turkey will also help in clearing the uncertainties about environmental restrictions on mining, so miners and investors will have clearer idea of what is permitted and what is not permitted. This will encourage companies to be more active in mining.

Madenköy (Siirt, SE. Turkey) copper deposit, which was discovered back in 1977 with the reserve of 27 million tons at 2.22% Cu, 32.2% S grades is to be exploited.

Preparations for the mining of Beypazarý (Ankara, Central Turkey) trona deposit (natural soda) have just started. Mining of soda will soon start. It is expected that new deposits will be discovered with the start of mining.

Boron deposits are very important for Turkey's economy. At present 9 by-products are produced and marketed. Turkey will increase the number of the boron by-products.

Payas (Ýskenderun, SE. Turkey) and Yalvaç (Ýsparta SW. Turkey) iron bauxite deposits; low-grade Hasançelebi (Malatya, E. Turkey) iron ore deposit; Uludađ (Bursa, NW. Turkey) wolfram deposit; Beylikahýr (Beypazarý, Ankara, Central Turkey) Thorium-fluorite-rare earth deposit all have some technological problems. When these problems are overcome, no doubt these deposits will be exploited.

Chromite deposits in Turkey are of alpine type. Exploration for new chromitite bodies generally goes along with mining. Careful examination and evaluation of geological data gathered during the process of mining have usually led to new discoveries. So far Turkey has not had any difficulties in meeting the market's demands.

Turkey has no experience and tradition of large-scale open-pit mining operations. Aladađ (Adana, S. Turkey) low-grade chromite deposit with the reserve of 200 million tons at 5.42% Cr_2O_3 grade is not operating (The reserve could go up to 400 million tons). Technological tests showed that the ore could be upgraded with 70 percent recovery. With suitable market conditions Aladađ chromite deposit will no doubt come into operation.

With the development of construction technology demand for light construction materials and for the materials, which provide heat and sound insulation will increase. It means that demand for perlite, pumice, gypsum, expendable clay will increase.

Marble quarrying and technology is one of the rapidly developing fields in Turkey. This branch of industry is expected to continue its development in the years to come.

Turkey has large areas of peridotite, but the use of forsteritic olivine in the refractory, moulding and heat insulation fields is very limited. It is expected that in future the use of olivine will increase, which is certain to cause the increase of olivine exploitation.

With favourable market conditions porphyry copper mineralizations will probably be operational.

5. Appendix

TABLE IV .

Metallic Minerals		
Mineral	Reserves	
	Reserve (mt) Proved+Probable	Grade-Quality
1	2	3
Gold	300	Metallic Au
Antimony	106.306	Metallic Sb
Copper	2.279.210	Metallic Cu
Bauxite	87.375.000	42-60 % Al_2O_3
Mercury	3.820	Metallic Hg
Zinc	2.294.479	Metallic Zn
Iron	149.925.000	55 % Fe
Silver	6.062	Metallic Ag
Chromite	25.931.000	20-50 % Cr
Lead	860.387	Pb content
Manganese	4.500.000	35 % Mn
Tungsten	36.719	Metallic W
Industrial minerals		
Alunite	4.000.000	7.54 % K_2O
Asbestos	29.646.379	In variable fibre length, fibre content over 4 %
Barite	35.001.304	71-99 % $BaSO_4$
Bentonite	250 543 000	Drilling mud additive, moulding sand binder, bleaching earth
Boron Minerals	1.805.709.953	24.4-35 % B_2O_3
Diatomite	44.224.029	Good quality
Cyanite	3.840.000	21-52 % Al_2O_3
Dolomite	15.887.160.000	Over 15 MgO
Feldspar	239.305.500	Albite and orthoclase
Phosphate	70.500.000	19 % P_2O_5
Fluorite	2.538.000	40-80 % CaF
Graphite	90.000	1-17 % C
Kaolin	89.063.770	15-37 % Al_2O_3
Rock Salt	5.733.708.017	Over 88.5 % NaCl (200 million tons of lake water reserve included)

1	2	3
Clay (ceramics & refractory)	354.362.650	Ceramics + refractory
Quartz Sand	1.307.414.000	Over 90 % SiO_2
Quartzite	2.270.288.000	Over 90 % SiO_2
Sulphur	626.000	32 % S
Meerschaum	1.483.000 (box)	Good + medium quality
Magnesite	111.368.020	41-48 % MgO
Perlite	5.690.028.000	Expansion ratio over 5 %
Pumice (m^3)	1.479.557.000	Good quality
Pyrophyllite	6.644.000	Ceramics + refractory + cement quality
Sepiolite	13.676.727	Sepiolite content over 50 %
Sodium Sulphate	16.536.000	81 % Na_2SO_4 (13.040.000 tons of lake water reserve included)
Celestite	665.072	Over 72 % $SrSO_4$
Talc	482 736	Good quality
Trona	233.317.680	Over 56 % trona
Zeolite	19.923.750	Clinoptilolite + heulandite
Emery	3.725.082	Over 50 % Al_2O_3
Energy raw materials		
Hard coal	1.126.493.000	
Lignite	8.257.268.000	

TABLE V a. Energy Raw Materials, Metallic / Non-metallic Minerals
Run-of-Mine (ton)

Minerals	Years					
	1995	1996	1997	1998	1999	2000 (9 Months)
1	2	3	4	5	6	7
Hard Coal	3 377 334(100)*	3 581 788(100)	-	3 336 021(100)	2 749 693(100)	2 389 808(100)
Lignite	56 031 099(91)	57 532 425(100)	-	66 583 499(96)	66 581 565(96)	49 379 618(96)
Asphaltite	147 154(100)	126 751(100)	-	157 334(100)	93 310(100)	76 339(100)
Iron	4 931 176(81)	6 279 951(84)	5 986 900(80)	5 885 429(82)	4 846 032(81)	2 761 743(81)
Chrome	2 080.043(16)	1 279 032(27)	1 646 413(29)	1 440 470(49)	1 019 542(53)	383 835(50)
Copper	2 927 864(100)	3 518 754(100)	3 794 630(100)	4 052 175(100)	4 310 485(100)	3 457 068(100)
Bauxite	232278(100)	544 513(100)	369 482(100)	458 028(100)	207 743(100)	396 667(100)
Antimony	7 856(100)	5 384(100)	585(100)	12 900(5)	13 991(0)	No Data
Manganese	37 000(100)	-	22 000(100)	53 283(79)	29 000(100)	5 000(100)
Zinc	73.110(9)	104 819(27)	79 500(5)	45 795(8)	4 630(0)	No Data
Lead-Zinc	240 862(40)	249 891(35)	262 260(0)	292 065(0)	284 504(0)	255 126(0)
Silver	640 000(100)	560 000(100)	660 000(100)	828 603(100)	650 000(100)	809 890(100)
Marble (m^3)	701 870(3)	1 190 520(2)	1 033 145(2)	1 583 580(2)	1 831 475(1)	496 263(1.53)
Building Stone	12 109 677(49)	11 587 724(75)	16 266 332(56)	-	-	No Data
Granite						
(Pegmatite)	-	251 352(10)	163 521(0)	18 198(0)	47 261(0)	20 008(0)
Basalt	193 996(100)	346 882(100)	519 500(100)	-	-	No Data

1	2	3	4	5	6	7
Limestone	31 174 131(10)	40 457 677(3)	49 685 582(7)	34 448 477(2)	27 633 895(3)	21 878 427(2)
Marl	4 156 543(6)	5 137 165(0)	6 150 726(1)	4 925 964(7)	3 433 329(0)	2 636 862(0)
Trass	1 622 527(1)	1 473 204(8)	2 447 800(24)	174 117(0)	99 114(0)	No Data
Plaster Stone	670 208(0)	754 277(1)	524 975(0)	351 557(0)	255 274(0)	193 080(0)
Dolomite	442 544(28)	795 695(29)	670 704(25)	829 775(31)	921 105(21)	555 292(26)
Sand and Gravel	33 570 778(12)	71 605 494(12)	86 120 647(10)	-	-	No Data
Silica Sand	737 605(13)	1 514 451(7)	899 853(0)	1 106 696(2)	1 225 399(1)	1 162 843(1)
Kaolin	489 635(3)	449 559(4)	1 099 340(2)	403 733(3)	449 954(2)	269 628(2)
Clay	3 958 319(54)	6 405 858(29)	5 932 656(23)	1 955 300(2)	2 357 208(2)	1 684 976(3)
Bentonite	605 499(0)	482 057(0)	521 036(0)	565 708(0)	1 051 995(15)	523 498(0)
Flintstone	-	-	13 047(0)	12 444(0)	12 444(0)	No Data
Marble Fragments	282 604(0)	279 824(0)	443 004(0)	576 928(0)	674 470(0)	549 240(0)
Stone Fragments	19 052 098(38)	17 013 584(46)	29 967 330(38)	-	-	No Data
Construction Filling Material	280 700(100)	1 886 516(90)	3 595 837(76)	-	-	No Data
Soil	885 212(0)	5 008 858(0)	4 049 613(0)	-	-	No Data
Boron Minerals	1 768 919(100)	2 400 635(100)	2 602 386(100)	2 754 081(100)	2 413 276(100)	1 560 809(100)
Pyrite	307 992(100)	538 140(100)	559 500(100)	699 408(100)	896 519(100)	255 914(100)
Barite	100 751(16)	104 872(16)	179 393(36)	160 042(29)	150 058(25)	89 977(9)
Fluorite	8 873(0)	4 828(0)	3 903(15)	5 034(21)	3 270(0)	2 978(28)
Sodium Sulphate	258 796(0)	369 867(0)	437 999(0)	412 247(0)	388 668(0)	289 599(0)

1	2	3	4	5	6	7
Strontium	-	-	-	1 673(0)	117 935(0)	No Data
Alunite	9 763(100)	6 625(100)	8 323(100)	10 624(100)	11 264(100)	9 382(100)
Zeolite	8 150(100)	-	-	2 470(0)	2 848(0)	No Data
Magnesite	1 928 064(19)	2 341 922(64)	2 050 818(6)	2 703 720(0)	-	1 933 525(0)
Rock Salt	75 671(100)	55 924(100)	64 889(100)	93 931(100)	93 931(100)	64 246(100)
Lake Salt	888 444(100)	1 386 262(100)	1 693 402(100)	1 618 000(100)	1 618 000(100)	599 997(100)
Sea Salt	464 089(100)	519 866(100)	532 202(100)	438 864(100)	438 864(100)	261 634(100)
Graphite	-	-	-	1 609(0)	2 078(0)	No Data
Emery	14 149(0)	-	12 345(1)	19 027(0)	14 535(0)	9 573(0)
Chalcedony	-	-	-	3 243(0)	3 102(0)	1 085(0)
Pumice (m^3)	1 125 820(25)	1 238 025(22)	935 578(25)	772 440(18)	986 386(15)	620 941(24)
Calcite	112 574(0)	129 062(0)	131 330(0)	138 619(0)	105 295(0)	90 574(0)
Perlite	171 058(7)	157 580(9)	103 416(14)	124 312(0)	142 061(12)	114 418(0)
Talc	-	-	-	894(0)	115(0)	107(0)
Diatomite	-	-	-	3 643(0)	5 926(0)	No Data
Feldspar	760 250(0)	912 204(0)	850 912(0)	1 089 483(0)	1 369 734(0)	944 389(0)
Quartzite	1 364 558(10)	2 511 565(5)	1 936 772(5)	2 301 270(5)	2 514 383(4)	2 291 854(2)
Nepheline-Syenite	-	-	-	56 391(0)	54 006(0)	No Data
Propylite	-	-	-	37 000(0)	46 739(0)	41 841(0)
Sepiolite	-	4 051(0)	9 083(0)	7 213(0)	7 718(0)	11 799(0)
Olivine	-	-	-	300(0)	300(0)	No Data

Source: DÝE (State Statistical Institute)
* Figures in brackets indicate the share of public organizations.

TABLE V- b. Oil and Natural Gas Productions

Products	Years					
	1995	1996	1997	1998	1999	2000 (9 Months)
Oil (Thousand ton)	3 516	3 500	3 457	3 224	-	2 068(73)
Natural Gas (Million m^3)	182	206	253	565	-	493(100)

TABLE VI. Minerals production of Turkey
(Million U.S. Dollars)

Oil-Natural Gas	500
Coal	1.500
Iron Ore	120
Copper	85
Chromium-Ferrochrome	160
Aluminium	125
Zinc-Lead	50
Perlite-Pumice	45
Boron Salts	270
Marble	350
Clay-Kaolinite-Bentonite	60
Magnesite	45
Quartz Group	70
Other Minerals	100
Lime-Chalk	300
Building Stones	800
Sand-Gravel-Fine Gravel	800
Other Building Materials	120
Total	**5.100**

TABLE VII. Share of mining in Turkey's export
(Million $)

Years	Total Export	Total Minerals Export	None-Metallics	Metallics	Energy	Mining Share in Export %
1990	12.959	327	242	82	1	2.52
1991	13.594	286	227	55	3	2.10
1992	14.714	264	212	49	1	1.79
1993	15.345	239	201	33	2	1.96
1994	18.105	272	215	50	2	1.50
1995	21.637	406	249	144	1	1.88
1996	23.123	389	253	118	1	1.68
1997	26.261	424	261	148	2	1.61
1998	26.973	385	256	111	4	1.43
1999	26.588	399	272	112	6	1.50
2000	27.324					

TABLE VIII. Share of mining in Turkey's import
(Million $)*

Years	Total Import	Total Minerals Import	None-Metallics	Metallics	Energy	Mining Share in Import %
1990	22.302	3989	91	82	3816	0.7
1991	21.047	2991	87	111	2793	0.9
1992	22.871	3054	80	80	2894	0.7
1993	29.428	3041	84	68	2887	0.5
1994	23.270	2969	77	55	2834	0.6
1995	35.709	3477	113	121	3238	0.6
1996	42.734	4297	115	138	4039	0.6
1997	48.559	4137	159	147	3821	0.6
1998	45.922	2942	153	176	2605	0.7
1999	40.686	3342	126	112	3100	0.6

* Energy (coal, natural gas, oil) not included.

TABLE IX. Share of mining in gross national product (GNP) of Turkey

Years	GNP 10^9 $	Mining GNP 10^9 U.S.$	Share of mining in GNP (%)
1990	152.351	2.405	1.58
1991	152.169	2.342	1.54
1992	160.688	2.176	1.35
1993	181.823	1.986	1.09
1994	130.888	1.821	1.39
1995	171.861	2.156	1.25
1996	184.602	2.256	1.21
1997	194.107	2.224	1.14
1998	205.808	2.166	1.05
1999	187.371	2.116	1.12

TABLE X. Precious metal (Au, Ag) deposits in Turkey

Deposit	Grade (gr/t)		Reserve (ton)	Metal content (ton)	
	Au	Ag		Au	Ag
Ýzmir, Bergama Ovacýk, W. Turkey	9.00	11.00	2 980 000	26.82	32.78
Ýzmir,Seferihisar,Efemçukuru,W.Turkey	12.65	-	2 500 000	31.62	-
Uţak,Eţme,Kýţladađ,W.Turkey	1.43	-	74 000 000	105.80	-
Balýkesir,Havran,Küçükdere,W.Turkey	6.43	11.80	1 410 000	9.07	16.64
Eskişehir,Sivrihisar,Kaymaz,C.Turkey	6.04	5.30	974 000	5.88	5.17
Gümüşhane,Mescitli,Mastra,E.Turkey	12.00	-	1 000 000	12.00	-
Çanakkale,Kirazlý,Akbaba,NW.Turkey	1.25	-	8 000 000	10.00	-
Artvin,Cerattepe,NW.Turkey (iron cap)	4.00	140.00	8 200 000	32 80	1 148.00
Massive Sulphide	1.20	25.00	3 900 000	4.68	97.50
Kütahya,Gümüşköy,NW.Turkey*	0.20	180.00	21 500 000	4.00	3 800.00
Total				244.00	5 100.00

* Operating silver deposit

TABLE XI. Mineral exports

Minerals	Years												% Change (1998/99)	
	1994		1995		1996		1997		1998		1999			
	Ton	1000$	Ton	1000$	Ton	1000$	Ton	1000$	Ton	1000$	Ton	1000$	Ton	1000$
1	2	3	4	5	6	7	8	9	10	11	12	13	14	15
Colemanite	326.788	75.303	347.219	81.086	374.191	84.622	364.852	79.963	338.509	74.041	339.930	72.397	0	-2
Tincal	145.586	23.492	164.335	26.460	176.587	28.476	181.837	28.712	148.708	23.914	106.800	17.818	-28	-25
Oulexite	149.821	22.077	170.150	26.148	185.100	27.922	185.030	27.449	168.300	25.886	206.750	31.104	23	20
Sinter magnesite	124.135	34.379	120.342	36.097	96.607	26.350	156.435	35.266	147.090	31.683	206.750	31.104	41	-2
Calcine magnesite	14.602	3.094	23.741	4.583	22.388	4.019	17.561	2.415	24.433	3.330	21.522	2.986	-12	-10
Magnesite (run-of-mine)	53.459	2.479	75.748	3.454	52.876	2.917	27.115	1.359	39.135	2.138	36.089	2.346	-8	10
Pumice	85.651	9.430	115.895	9.135	145.880	10.796	88.066	8.073	94.966	8.517	117.064	6.394	23	-25
Barite	63.693	4.109	104.991	6.367	124.687	8.079	125.411	8.713	123.972	8.163	57.556	3.584	-54	-56
Plate marble	16.646	3.186	10.798	2.132	15.828	2.153	23.018	3.602	21.474	5.068	30.949	6.152	44	21
Block marble	44.184	5.959	57.660	7.852	48.646	7.593	79.738	11.029	91.860	12.096	130.361	22.998	42	90
Perlite	131.812	2.897	202.029	4.601	161.526	4.212	178.054	3.942	141.897	3.874	145.060	3.448	2	-11
Celestite	22.000	1.496	23.520	1.600	49.350	3.405	46.110	3.228	29.570	1.981	60.540	4.090	105	106
Kaolin-Clay	189.423	4.137	256.907	5.892	188.666	5.150	158.642	3.878	185.431	5.592	262.211	7.464	41	33
Bentonite	53.693	2.436	54.315	2.968	48.528	2.431	68.718	2.790	90.182	4.133	104.425	4.619	16	12
Feldspar	469.370	9.505	730.937	15.445	770.171	15.933	950.140	20.692	1.283.090	24.500	1.799.853	32.735	40	34
Salt	4.399	336	10.271	575	10.377	692	13.116	981	17.427	864	20.783	1.225	19	42
Other	145.453	16.592	288.762	14.655	367.364	18.444	271.127	18.752	344.028	19.516	341.024	22.009	-1	13
Non-metallics total	**2.040.715**	**220.907**	**2.757.620**	**249.050**	**2.838.772**	**253.194**	**2.934.970**	**260.844**	**3.290.072**	**255.296**	**3.987.667**	**272.473**	**21**	**7**

1	2	3	4	5	6	7	8	9	10	11	12	13	14	15
Chromite concentrate	596.100	33.151	1.000.362	91.859	519.299	59.124	588.564	57.495	495.136	45.638	559.425	38.606	13	-15
Copper ore and concentrate	41.233	10.027	83.834	35.024	111.435	28.740	180.669	49.732	154.149	33.196	194.300	44.326	26	31
Zinc ore and concentrate	17.332	3.665	69.954	13.331	107.672	21.907	108.186	31.138	101.840	22.487	100.503	23.929	-1	6
Aluminium ore and concentrate	30.160	778	15.350	430	21.000	357	110	35	226	64	130	23	-42	-64
Chrome ore	6.000	388	12.940	1.355	8.398	960	35.350	3.467	42.516	3.382	7.161	661	-83	-80
Other metallic ores	148.690	6.755	154.139	13.166	387.398	23.730	160.303	18.874	148.433	19.793	101.783	12.751	-31	-36
Metallics total	**839.515**	**54.776**	**1.336.579**	**155.165**	**1.155.202**	**134.818**	**1.073.182**	**160.741**	**942.300**	**125.280**	**963.302**	**120.296**	**2**	**-4**
Min.s+Metallics total	**2.880.210**	**275.683**	**4.094.199**	**404.215**	**3.993.974**	**388.012**	**4.008.152**	**421.585**	**4.232.372**	**380.576**	**4.950.969**	**392.769**	**17**	**3**
Processed marble	105.063	50.755	129.830	63.468	169.927	79.142	206.183	94.305	234.191	104.990	292.642	122.611	25	8
Minerals total	**2.985.293**	**326.438**	**4.224.029**	**467.683**	**4.163.901**	**467.154**	**4.214.335**	**515.890**	**4.466.563**	**485.566**	**5.243.411**	**515.380**	**17**	**8**

TABLE XII. Mineral imports

Minerals	Years								% Change	
	1996		1997		1998		1999		(1998/1999)	
	Ton	1000$	Ton	1000$	Ton	1000$	Ton	1000$	Ton	1000$
Phosphate	645.805	22.067	986.834	42.462	1.041.910	45.031	873.874	39.084	-16	-13
Kaolin	83.690	13.089	113.432	15.346	139.326	18.653	114.965	13.430	-17	-28
Plate marble	27.819	5.001	36.300	7.955	36.050	8.391	462	67	-99	-99
Salt	3.511	735	4.660	1.462	7.643	3.363	10.707	3.031	40	-10
Sulphur	115.355	7.494	170.473	9.851	126.446	10.030	133.554	7.603	6	-24
Quartz	174.413	6.305	168.628	6.776	213.249	8.032	1.046	252	-100	-97
Asbestos	26.868	12.460	33.823	15.797	22.674	9.687	5.618	15.494	-75	60
Feldspar	18.034	3.064	19.802	3.983	26.240	3.286	633	239	-98	-93
Others	115.183	45.519	177.790	56.598	143.417	46.983	463.535	46.541	223	-1
Non-metallocs total	**1.210.678**	**115.734**	**1.711.742**	**160.230**	**1.756.955**	**153.456**	**1.604.394**	**125.740**	**-9**	**-18**
Iron ore	2.991.252	115.971	2.950.732	106.419	3.857.734	141.774	2.973.052	89.468	-23	-37
Zinc ore and concentrate	31.613	8.681	68.120	13.448	83.377	18.942	59.099	6.779	-29	-64
Copper ore and concentrate	735	4.310	35.952	18.685	8.218	2.283	18.369	6.320	124	177
Chrome ore and concentrate	5.203	813	822	203	19.497	4.871	48.284	4.013	148	-18
Aluminium ore and concentrate	34.874	4.527	25.680	4.145	24.844	4.550	25.247	3.018	2	-34
Other	72.672	5.214	247.161	10.618	304.202	8.729	62.429	5.235	-79	-40
Metallic total	**3.136.349**	**139.516**	**3.328.467**	**153.518**	**4.297.872**	**181.149**	**3.186.480**	**114.833**	**-26**	**-37**
Processed marble	17.250	13.395	24.259	20.436	41.406	34.871	42.415	28.973	2	-17
Mineral total	**4.364.277**	**268.645**	**5.064.468**	**334.184**	**6.096.233**	**369.476**	**4.833.289**	**269.546**	**-21**	**-27**

MINERAL RESOURCE BASE OF AZERBAIJAN: A GAZE INTO THE XXI CENTURY

V.M. BABA-ZADEH, G.V. MUSTAFAYEV, V.G. RAMAZANOV

1. Introduction

Azerbaijan, as the biggest country in the South Caucasus and a transit zone between Europe and Asia, possesses considerable mineral resources — hydrocarbonic as well as solid ones.

But historically it so happened that the discovery more than one and a half century ago of high-grade oil fields on the Apsheron peninsula, and later on other sites, not only on the firm-land, but also in the adjacent water areas of the Caspian Sea, made Azerbaijan into the world's most important oil region.

In 1949 there was discovered the first offshore oil field of «Neft Dashlary» («Petroleum stones»), which has become world-famous by now. The subsequently revealed major offshore oil fields of Azeri, Chirag and Gyunashli, laying the foundation of the «Contract of the century», concluded in 1994 between Azerbaijan and the world's largest oil companies and consortia, on the verge of the beginning of full-scale exploitation whereof we are today, and also the oil pool of Kyapaz, are situated on the continuation of this very same unique oil field.

Nowadays 20 petroleum contracts with the aggregate volume of investment of about 60 billion dollars are signed with oil companies of 15 countries. There is discovered the largest offshore gas condensate field of «Shakhdeniz», having reserves of natural gas of more than 1 trillion cubic metres. The discovery of «Apsheron» gas field, more considerable than Shakhdeniz, is prognosticated.

However, the fact that Azerbaijan is doubtless a country of great interest in respect of ore mineral resources, was later demonstratively proved by amplitudinous geological prospecting, especially in the second half of the XX century, which allowed to regard the Azerbaijani portions of Greater and Lesser Caucasus as potentially promising with respect to different kinds of solid mineral resources, including bulky ones. With the discovery in 1958 of the unique Filizchai pyrite-polymetallic deposit and thereafter of the entire group of analogous industrially valuable ore sites (Katsdag, Katekh, Sagator, Djikhikh, Mazymchai, and others) there took place the fundamental change of the whole metallogenic content of the entire south-eastern segment of the Greater Caucasus.

The prospectiveness of the metallogenic province of the Greater Caucasus is further enhanced with the revelation of auriferous manifestations of «black schist» formation in the potent zones of secondary schistosity of linear folds of Tfan and Sarybash structural-

A.G. Tvalchrelidze and G. Morizot (eds.),
Mineral Resource Base of the Southern Caucasus and Systems for its Management in the XXI Century, 105–124.

formational zones (Gyzylgai, Aglyg-Filfilli, Vandam-Galadjik and others).
The ore characteristic of the Lesser Caucasus has been substantially changed both by the re-evaluation of previously known iron ore, copper pyrite, barytes and other deposits, and by the revelation of new, especially copper-porphyry, gold ore industrially important deposits, a major deposit of zeolites, etc. There is established the regional placer auriferosity within the Lesser Caucasus, including its vast piedmont area.
There are obtained significant data concerning ore bearing capacity of Upland Talysh.
So, today Azerbaijan, alongside with being a generally recognized centre of hydrocarbonic mineral resources, is a country possessing huge reserves of solid mineral resources, whose «face» is certainly determined by iron, aluminium (alunite), copper, lead and zinc, gold, mercury, barytes,
zeolite, and others, to say nothing of numerous occurrences of various mineral resources, used by ferrous metallurgy, chemical industry, occurrences of mineral building materials and basic materials for their production, etc. Azerbaijan's subsoil is also rich in different kinds of mineral, thermal and subsurface industrial (iodine-bromine) waters, whose reserves are really enormous.
The territory of the republic of Azerbaijan and adjacent water areas of the Caspian Sea embraces from north to east parts of the frontal flexure of the Caucasian foreland, of the plicate structure of the Greater Caucasus, of the Kura intermontane flexure, of the Lesser Caucasian plicate structure, and of the rift zone of Talysh. The formation of the endogenous ore fields of the Azerbaijani segment of Caucasus, entering into the composition of the global Mediterranean metallogenic belt, is due to the Alpine metallogenic epoch, comprising the period from the Early Jura to the time of Anthropogeny and resolving into the Early Alpine and the Late Alpine stages.
In the Alpine metallogenic epoch in the Greater and Lesser Caucasus within structural-formational zones there can be distinguished local narrow geochronological levels of maximum accumulation of mineral resources.

2. Mineral Resources for Ferrous Metallurgy

2.1. IRON

The Dashkesan deposit is one of the major iron ore sites in the former Soviet Union. Commercial reserves amount to more than 300 million tonnes, which allows to meet the needs of Azerbaijani Metal Enrichment Plant, with the output existing by the year 1990, approximately for one hundred years. Magnetite skarn mineralization is represented by the gently dipping stratiform deposits, extending in some areas (north-western, north-eastern, and others) for 1900 to 4000 metres. Lenticular-stratiform ore bodies have the extent ranging from 100-200 to 1500-2000 metres with the maximum depth of up to 56 metres.
The iron content in massive ores amounts to 45-60%. The vanadium and cobalt content (up to 0.04%) in ores is sufficiently high.

In the north part of the Dashkesan iron ore deposit there is known a separate plutonogenic hydrothermal deposit of cobalt (Dashkesan-cobalt).

2.2. MANGANESE

Comparatively well explored is a number of smaller but promising deposits of Late Cretaceous (Upper Santonian) age, pertaining to the volcanogenic-sedimentary genesis (Molla-Djalily, Elvor, Dashsalakhly, Dalidash-Djogaz, Karayatag, and others). The sheet deposit of Molla-Djalily occurrence ranging in extent from several tens of metres to 50-90 metres, comes up in the central section of the deposit to 200 metres with the depth of 0.3-3.0 metres, occasionally 10 metres. The manganese content amounts to 17-32%, the iron content – to 20.4 – 42%, the P_2O_5 content – to 1.54%.

3. Mineral Resources for Non-Ferrous Metallurgy

3.1. ALUMINIUM

It is being extracted from alunite ores of the major Zaglik deposit. Taking into consideration the generous commercial reserves of the latter (130 million tonnes), exploratory operations for these ores are suspended. A contract is sealed with the Dutch company «Fondel metall Participatione B.V.» on transferring «Azerbaijani aluminium» for long-term management. Presently prospecting operations are mainly concentrated on looking for more economic bauxite ores, whose manifestations are found in the south of the Lesser Caucasus in lower parts of Permian sediments, deposited upon various horizons of the Lower Carboniferous Period (Gerangalasy, Gabakhyal, Sadarak, Danzik, Kyarky, and others). The thickness of bauxites (silicon module above 2.1) and bauxite rocks amounts to 2-13 metres and can be traced on the areas of revealed ore manifestations down to the depth of 1.5-2.0 kilometres, forming stratiform-lenticular bodies. The Al_2O_3 content in bauxites amounts to 30-57%.

3.2. COPPER AND MOLYBDENUM

Commercial deposits of copper and molybdenum are mainly known in Kedabek and Ordubad ore regions. They are represented chiefly by two commercial and genetic types — by copper- and molybdenum-porphyry (prevailing) (Karadag, Kharkhar, Djagirchai, and other deposits in the Kedabek ore region and Misdag, Geidag, Diakhchai, Geigel, and other deposits in the Ordubad ore region) and by copper pyrite (Kedabek, Bitti-Bulakh and other deposits) ones.

Within the Kharkhar area of copper-porphyry ores there is revealed the homonymous deposit of turquoise. Reserves of the latter by C_2 category amount to 2 tonnes. The raw material is chiefly of the first- and second-rate quality.

Mentionable are the deposits and manifestations of the cupreous sandstones of the

Asadkaf group (Khalkhal, Sirab, Yaidji, Kagat, and others), extending along the south-east edge of the Nakchyvan superimposed trough and timed to the volcanogenic-sedimentary layer of Oligocene, controlled by the regional Nakchyvan rift of south-east extension. The layer forms a series of brachyantyclinal stuctures, corresponding to the volcanic centres and extrusive tumuli of andesite-dacite composition. Mineralization has stratified, but discontinuous character and is revealed on three horizons with the depth of 3-5 metres within the area of development of Oligocene sediments having the total extent of 60-70 kilometres. Extent of some mineralized areas, of which there are revealed twenty, reaches 1-1.5 kilometres. The copper content in cupriferous strata averages to 0.3-0.5%, in beneficated ones — to 0.5 to 1.56%.

Undoubtedly, the plutonogenic hydrothermal deposits of fissure-vein type (Paragchai), combined in their lower horizons with the copper-porphyry ones, have encouraging prospects.

3.3. LEAD AND ZINC

In the Arax zone in the south of the Lesser Caucasus there is located a separate Gymushlug ore field with the homonymous deposit and prospective ore manifestations of Danzik, Sadarak, and others, timed to the limestones of the Givetian assise. Within the Gyumushlug ore field multistage mineralization is possible. Taking into consideration the presence on the territory of the conterminous side of Iran of high-grade deposits of polymetallic ores with increased lead content (up to 13%), it could be possible under analogous geological conditions both within the Gyumushlug ore field and in the area of essentially carbonaceous sediments of Palaeozoic to discover latent deposits of massive stratiform lead-and-zinc ores.

In the exocontact of the Megri-Ordubad batholite in the spatial connection with volcano-tumulous structures of dacites and rhyolite-dacites there is discovered a large number of prospective lead-zinc deposits (Agdara, Nasirvaz, Kvanuts, Kovurmadarasi, Mazri, Sapardara, and others).

In the Somkhit-Agdam zone the Mekhmani deposit of the Middle Jurassic (Bajosian-Batonian) age is considered to be the most promising lead-zinc ore field. It is a typical vein hydrothermal deposit, displaying close spatial connection with minor diorite-porphyry intrusions.

Within the country there has been revealed more than 20 occurrences of pyrite-polymetallic ores. Most of them belong to the stratiform geological-commercial type and are located within the Belokan-Zakatala ore region, comprising the well-known Filizchai deposit, which is unique by the summary amount of the calculated reserves of mineral raw materials, for with regard to zinc, lead, cadmium, sulphur and particularly to noble metals it is considered to be large. In the ores of the Filizchai deposit there also participate with commercial significance bismuth, selenium, tellur, cobalt, and others. Apart form Filizchai, in the ore region there are also Katsdag, Katekh, Mazymchai, Sagator, and other deposits, which are characterized by large explored reserves. There is a real prospect of increasing the reserves both at the expense of the less thoroughly investigated objects (Gyumbulchai, Katsmali, Batoor, and other deposits), and by exploring flanks, contiguous areas and deep horizons of the already known deposits.

The pyrite-polymetallic deposit of the Filizchai field is represented by the aggregate stratiform body. By its strike the explored portion of the ore deposit amounts to 1800 metres, by its dip — to 1500-1600 metres. The thickness of the deposit consistently and gradually increases from west to east from 0.1-0.2 metres to 25 metres. With the presence of bulges and twitches of the ore deposit (from 0.5 to 60.01 metres), its different parts vary notably from one another in details of their composition. Reserves of schistose-banded and massive structure have sharp and distinct contacts with the enclosing rocks both on the side of the bottom and that of the top. Entire masses of copper-pyrrhotine ores also participate in the composition of the deposit, forming 2% of the total amount of ores.

There can be observed the pyrite ores being strictly timed to the definite stratigraphic levels of the Lower Middle Jurassic complex (Plinsbachian, Toarian and Upper Aalenian). The fact of the ore deposits being of stratified and multistage character considerably increases in its turn the prospects of revealing the ores of the Filizchai type in the deep horizons.

Large reserves of pyrite-polymetallic ores of the south slope, potential possibility of their manifold increase, considerably enhance the prospects of Azerbaijan's north-east part and entitle us to maintain the real possibility and economic advisability of establishing here, on the basis of advanced technology, a major centre of non-ferrous metallurgy with an integrated polymetallic plant for the complex extraction of all useful components, contained in the ores of the Filizchai and other deposits, the possibility to exclude losses in the bowels and tailings of the flotation of copious reserves of sulphur-ore, suitable for the production of sulphuric acid and elementary sulphur.

3.4. MERCURY, ARSENIC, ANTIMONY

Within the confines of the Geicha-Karabakh zone, especially on its Terterchai sector and the south-east flank, there is discovered more than ten practically monometallic mercury (Agyatag, Shorbulakh, Agkaya and others), antimony (Elizgel), antimony-mercury (Levchai) fields, increasing in complex the prospects of the zone for the metals in question. The listvenite type of mercury deposits is widely spread.

In the Arax zone there are known the Darrygadag (Djulfa), Salvarti, Ortakend, Bashkend fields and the manifestations of arsenic ores. In the zone of the Darrydag ore-bearing rift there is supposed the presence at a depth chiefly in the Palaeocene sand bands (from 2-3 to 6-7 metres thick) of commercial accumulations of these metals, which is confirmed by the primary mineralization of arsenic in association with its primary aureoles in combination with antimony, copper, rarely with zinc, etc., connected with the sand bands on the Dagustu area. The fact of high arsenic contents being timed to one and the same horizon testifies to the great importance of the lithological factor for the disposition of mineralization, which increases the reliability of the prognosis made.

Comparatively recently revealed Ailis deposit on the south-west slope of Mt. Arysu in the Ordubad ore region is of considerable interest.
Mineralization of antimony, being timed to the casings of steeply dipping dikes of diorite-porphyrites, is localized in the enclosing arenaceous-argillaceous and pyroclastic formations (Arysu suite — Upper Pliocene-Lower Eocene). The extent of ore bodies is 200-800 metres, throughout the zone it reaches 1700 metres, average thickness is 7.0 metres. The basic ore mineral is antimonite. Average content in tested sections amounts to 2.5-3.0%, in some sections to 6.5-7.3%, rarely to 13% (with the thickness of 3.0 metres). In antimony ores there have been also found arsenic, copper, more rarely gold and silver.

4. Noble Metals

4.1. GOLD

Gold ore objects of the Lesser Caucasus are investigated not with the equal grade of thoroughness. Only four fields (Kyzylbulakh, Vezhnali, Goshi, and Dagkesaman ones) which are most promising have passed the stage of detailed prospecting. In all of these deposits gold ore mineralization is concentrated in comparatively narrow and extended zones of hydrothermally altered rocks. In the said deposits commercial reserves are evaluated and they may be involved in being developed.
In the Ordubad ore region gold ore deposits and manifestations are grouped in two auriferous zones. The first one is located in the endocontact part of the Megri-Ordubad batholite and unites a series of assemblies with auriferous pyrite-chalcopyrite veins (Agyurt, Bashyurt, Munundara), the second one is timed to the strip of volcanic rocks enclosing batholite and containing quartz-pyrite and quartz-polymetallic veins, a number of which (Pyazbashi, Shakardara, Kyalyaki, Kvanuts, and others) is characterized by the commercial auriferousness. The special emphasis must be laid on the Pyazbashi deposit. Prospecting has proved the commercial value of the vein # 1, whose reserves are calculated. Vast number of auriferous quartz veins (more than 70) within the ore field, as well as numerous veins, revealed by prospecting operations, dictate the necessity to estimate their depths. It is necessary also to take into consideration the commercial value of the Munundara deposit, where reserves are calculated, technological tests of ores have been carried out and the scheme of their benefication is made up.
Important objects are represented by the group of essentially auriferous pyrite deposits, in the quantities, allowing to rate them among the main ore-forming components (Kedabek and others). Gold in the form of admixture is present in all deposits of the copper-porphyry formation of the Lesser Caucasus. Auriferous gravels of the Upper Quaternary alluvial deposits, especially the lowland portions of the rivers Akstafachai, Tauzchai, Kyurakchai, Asrikchai, Dzegamchai, Shamkirchai, Terterchai, Alindjachai, Vanandchai, Gilyanchai, and others are the objects of major practical interest. The signs of the molasses being auriferous are discovered in Nakchyvan and in a number of river basins of the north-east slope of the Lesser Caucasus.

5. Chemical Raw Materials

5.1. BORON

Efficient for the extraction of this valuable and scarce mineral raw material could be volcanic breccias of dirt volcanos, in which oulexite and borax are discovered.
According to the data obtained on 16 dirt volcanos alone, the calculated amount of boric anhydride (B_2O_3) in volcanic breccias is equal in a first approximation to about 7.5 million tonnes, with its average content of 0.16%. In the products of these effusions high contents are also established for lithium (0.0045-0.1%), rubidium (0.0008-0.025%) and caesium (0.0025-0.01%).
In the saliferous layers of Nakchyvan there is established the presence of hydroboracite and achardite, and in the contact strip of minor intrusions — of the manifestations of datolite (datolite skarns of Mt. Ilandag and others).
Considerable importance in respect of boron can be attached to the carbonaceous-arsenic springs of the Darrydag group (Nakchyvan), containing besides arsenic (15-21 mg/l) also increased concentrations of boric anhydride in the form of boric and meta-boric acids (800-1200 mg/l) with rather high yields.
The analysis of geological material attests to the presence of commercial concentrations of boron and high yields and pressures of the boron-bearing fissure rocks along the Darrydag zone of deeply bedded faults.
Thus, the question of commercial value of the boron-bearing springs of Nakhchyvan, as well as of the volcanic breccias of dirt volcanos, and of the saliferous layers of various structural-formational zones admits of no doubt. In this particular case commercial development of the major raw material base of liquid borates and the organization of the production of boron on the basis thereof is limited by the absence of a reliable technological scheme for the extraction of boron anhydride from arsenic waters.

5.2. DOLOMITES

They are stretched in a wide strip along the left bank in the middle reaches of the r. Arax. The thickness of these formations reaches 1200-1500 metres. On the whole extent the Upper Triassic dolomites are conformably bedded upon the faunistically characterized sediments of the Middle Triassic. The largest one is the Negram field (commercial reserves amount to 140 million tonnes), timed to the Djulfa anticlinorium of the second order. The efficient layer, forming a large sheet deposit of homogenous grey, pinkish-grey and pink dolomites has the thickness of 980 metres. They are eligible for their purity and homogeneity, no terrigenous-siltstone-arenaceous material has been found in any of the types. By their chemical composition and physical and mechanical properties the dolomites meet the requirements of the State Standard and may be used as a facing material (especially their pink varieties), and also as a high-strength road metal and constructional break-stone. They can be used as a refractory material of the I class, in the production of soda-ash and metallurgical magnesia and as glass stock.

5.3. SALT

5.3.1. *Rock-salt and self-precipitated salt*

The largest fields of rock-salt are those of Nakhchyvan, Susta, Duzdag, Negram, and Shakarabad-Koshadizi, forming, substantially, a major saliferous basin of the Neogene age. The Negram deposit is timed to the roof of the Middle Miocene sediments. In respect of its dip the rock-salt deposit is traced for 3 kilometres, in respect of its extension — for 400 to 800 metres. The depth of the deposit is 46.4-90 metres. The sodium chloride content amounts to 83-96%. By the pure magnesium content the salt conforms to the grades «extra» and «superior». The main factor deteriorating the quality of the salt is a high content of insoluble residue (1.38-8.96%). The quality of the salt improves sharply when developing the deposit by using the method of subsurface lixiviation. The object is of major practical interest for the chemical industry, not only for the production of soda-ash, but also for releasing brines of soda salt and self-precipitated salt, which will be used for the production of superior quality table salt and chemically pure salt. Prognosticated reserves amount to 1.2-1.5 billion tonnes.

5.3.2. *Self-precipitated salt*

Its deposits are known mainly on the Apsheron peninsula. The largest ones are Beyuk-Shor, Masazyr, Byulbyuli deposits.

5.4. BARYTES AND ZEOLITES

Being multipurpose raw materials, barytes and zeolites are the objects of indubitable interest. Prognosticated resources of the Chovdar deposit amount to more than 1.5 million tonnes, which is quite acceptable for organizing a mine. In the west of the country, in the Tauz region there is revealed a large Aidag deposit of Late Cretaceous (Upper Santonian-Lower Campanian) tuffs with a high natural zeolite-clinoptilolite (commercial reserves of 30 million tonnes) content (up to 70-90%, on the average 51.3%). In association with the latter there occur mordenite, heulandite, natrolite, analcite, and other minerals. The extent of the zeolite-bearing layer amounts to 1.5 kilometres, the depth — 35 to 100 metres. Ash tuffs of andesites, dacites, rhyodacites, etc. are in most cases the source material for the formation of zeolites. The deposit is structurally timed to the Kerogli synclinal fold and is located at the point of intersection of the rifts of north-west and north-east extension. Other deposits of the Kazakh downfold (Tatli, Kerogli, Gaimagli, Ali-Bairamli, and others) have approximately analogous structure.

Excellent sorptive properties and the good economical advantage of using the natural zeolites of the Aidag field make this kind of raw materials very precious and indispensable. Investigations carried out by various scientific and applied research institutes have proved the applicability of the natural zeolites of the Aidag deposit for the deep dehumidification and cleaning of natural gases of hydrogen sulphide, carbon dioxide, butane, and higher hydrocarbons, for the cleaning of gases of sulphate generation of sulphur oxides, for the extraction of rubidium, caesium, lithium, and strontium from natural and waste waters, for the separation of gas mixtures of oxygen, nitrogen, argon.

5.5. BENTONITE CLAYS

They are concentrated mainly on the north-east piedmont areas of the Lesser Caucasus in the volcanogenic-sedimentary formations of Cognac-Santonian, partly, of Campanian assises of the Somkhit-Agdam zone (Dashsalakhli, Khanlar, and others). In recent years there are revealed large enough deposits of high-quality bentonite clays of the Paleogene age (Shikhandag, Beglyar, Kyashkurak, and others) upon the south-eastern immersion of the Greater Caucasus.

The Dashsalakhli field in the Kazakh region (commercial reserves amount to 100 million tonnes) is represented by a stratiform deposit with two sections — the Central and the South ones. In the efficient layer there can be distinguished two strata of bentonite clays, formed at the expense of the hydrothermal transformation of tuffs. The bentonites consist of montmorillonite (70-85%) with the subordinate amount of kaolinite, hydromica, calcite, tridymite, less frequently — of metahalloysite and gypsum. The alkalinity coefficient of the bentonite clays is on the average 2.3 high, which allows to rate them among the alkaline varieties, turgescence in the air-dry condition varies from 10 to 19 times, chiefly 12-13 times. The highly turgescent (more than 12 times), highly dispersed clays are fit for making pellets of iron ore concentrates (more than 50% of commercial reserves), mud-flushes, etc.

5.6. WATER RESOURCES

5.6.1. *Underground commercial (iodine-bromine) waters*

In Azerbaijan there are located the highest-grade deposits of underground commercial waters with the reserves of 250 thousand m^3/day and with high iodine and bromine contents. Major reserves of iodine-bromine waters are concentrated in the Babazanan, Mishovdag, Khilli, Neftechali, Bina-Govsan, Koturdag-Kalmas, and other fields. The now functioning Baku and New-Neftechali iodine-bromine works make use of just a small part of the huge reserves of this mineral raw material. Underground commercial waters contain strontium, which is not recovered under the applied technology. That is why it is necessary to develop a more perfect technological scheme of production with the recovery of not only iodine and bromine, but also of strontium and other rare elements.

5.6.2. *Mineral waters*

Numerous deposits of table waters and healing waters have occurred in various zones of the country. It is enough to mention the well-known ones of them, namely those of Istisu, Badamly, Turshsu, Sirab, Vaikhyr, mineral water springs of the «Narzan» type in the Kedabek and Ordubad regions, and others. The utilization of this material wealth rests first of all on bottling companies being poorly provided with glass containers. Another, not less important aspect of the problem is the absence of health resort engineering, given such unique healing waters, like the «Naftusi» type Galaalty one in the Divichi region, the Darrydaga arsenic springs in Nakchyvan, the springs in the basin of the r. Tutkhun (Gotursu), and others, situated in the picturesque places of the country.

5.6.3. *Thermal waters*
In the Masally-Astara zone there is concentrated a considerable storage of high-temperature mineralized waters, which may be used for heating and in greenhouse-hotbed farming producing early vegetables. Thermal waters have been revealed also in the Kuba-Khachmaz zone.

5.6.4. *Underground waters*
There are explored and calculated the exploitation reserves of the underground waters fit for water supply, irrigated agriculture, etc. contained in the main artesian basins of the country, the largest of which are Alazan-Agrichai, Kuba-Mil, Samur-Velvelichai, Lenkoran, Djebrail, Shirvan ones, which has allowed the specialists to make up a scientifically established scheme of the utilization of underground waters.

6. Conclusion

As we know, after the collapse of the Soviet Union and the sovereignization of the republics the well-established economic ties between them were broken, appropriation of funds for prospecting operations was sharply cut down and as a result of this discoveries of new deposits decreased in number, and in many cases extraction of mineral products was reduced to nothing at all.
In order to avoid the further development of the said phenomena and to stabilize the operations of the country's mining industry, and also further to improve Azerbaijani economy, it is considered to be necessary to develop a new approach to the efforts to strengthen the country's mineral resource base. It is the main objective of the next few years, to develop the explored resource bases and, in prospect, the series of the revealed rather considerable deposits, satisfying the present-day requirements of market economy. Greatly conducive to this, as we believe, is the important geopolitical position occupied by Azerbaijan as a transit country, and the region's eligibility for foreign investments. Top-priority objects, which may play a pivotal role in the development of the country's mineral resource base, are the deposits with high quality and considerable reserves of ores and favourable mining conditions for their extraction. In this respect it is advisable:
— to construct in the north-west of the country on the basis of the Filizchai group of pyrite-polymetallic deposits the metal enrichment plants for the extraction and reduction of ores and for the production of hardware;

— to construct in the Gyandja-Kazakh economic zone a metallurgical plant on the basis of the Dashkesan iron ore deposit. Large reserves of iron ores and the presence in the said economic zone of practically all sorts of mineral raw materials, necessary for the metallurgical industry (high-quality fluxing limestones, alkaline bentonites fit for the pelletizing of iron-ore concentrates, refractory lining materials, silica brick stock, etc.), as well as the well-developed industrial infrastructure speak in favour of the advisability of organizing a metallurgical plant here;
— the great number of gold-ore and goldbearing deposits dictates the necessity to design and construct mining and concentrating plants and by doing so to lay the foundation of Azerbaijan's gold-mining industry.
— under certain conditions it seems possible to apply a group approach to the geologic-economical evaluation of the deposits of the same tenor, situated near one another, especially to copper- and molybdenum-porphyry ones, characterized in the aggregate by considerable reserves and being of indubitable commercial interest.
— to develop deposits of natural zeolites, particularly that of Aidag, which is one of the largest fields within the post-Soviet space, having rather high clinoptilolite content;
— to construct on the basis of the Negram rock-salt deposit in the Arax zone a major soda plant for the production of calcinated salt, soda salt and salt obtained by evaporation, as a secondary production of metallurgical magnesia and electrosmelting periclase;
— high concentrations of boron material in the volcanic breccias of Apsheron dirt volcanos and in the saliferous layers and boron-bearing springs of Nakhchyvan make their commercial development and the organization of boron production on their basis indispensable;
— to commercially develop iodine-bromine waters, having substantial reserves, using perfect technology for the extraction of not only iodine and bromine, but also of strontium and other rare elements;
— to make extensive use of various types of mineral waters, developed in different structural-formational zones, first and foremost — to organize resort engineering on their basis health.

7. Appendix

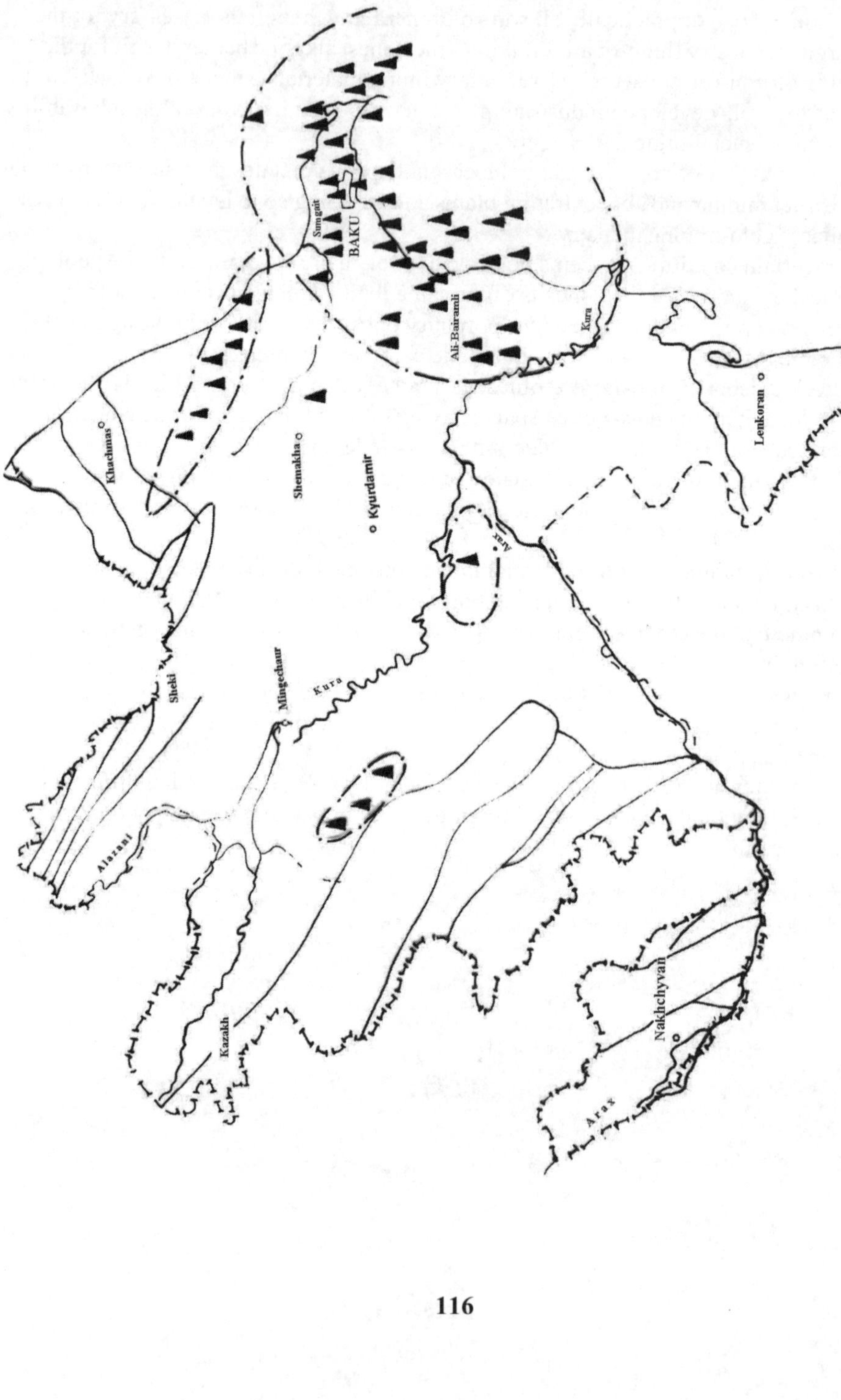

Figure 1. Chart of Location of Deposits of Mineral Resources of Azerbaijan Oil and Gas Deposits and Areas of their Distribution

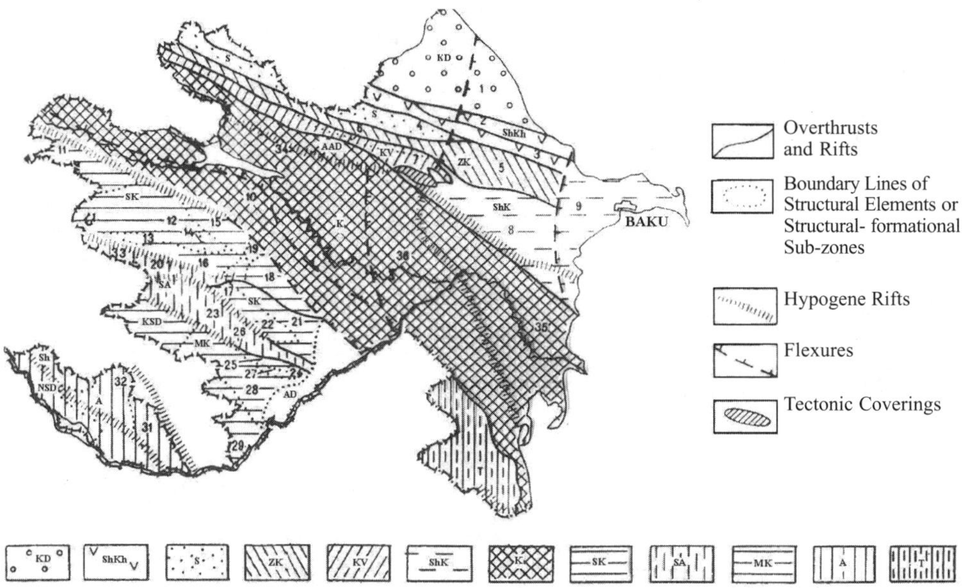

Megastructures	*Structural-formational Zones*	*Tectonic Structures*
Megaanticlinorium of the Greater Caucasus	Kusar-Divichi (KD)	Kusar-Divichi Downfold (1)
	Shakhdag-Khizi (ShKh)	Tengin-Beshbarmag Anticlinorium (2), Shakhdag-Khizi Synclinorium (3)
	Sarybash (S)	Tfan Anticlinorium (4)
	Zakatala-Kovdag (ZK)	Zakatala-Kovdag Synclinorium (5), Lagich Synclinorium (6)
	Kakheti-Vandam (KV)	Vandam Anticlinorium (7)
	Shemakha-Kobustan (ShK)	Shemakha-Kobustan Synclinorium (8), Apsheron Periclinal Downfold (9)
Kura Mega-synclinorium	Kura (K)	Alazan-Agrichai Downfold (AAD), Lesser Caucasian Foreland Downfold (10), Middle Kura Downfold (34), Lower Kura Downfold (35), Talysh-Vandam Buried Uplift of Periclinal Downfold (36)
Megaanticlinorium of the Lesser Caucasus	Somkhit-Karabakh (SK)	Kazakh Synclinorium (11), Shamkhor Anticlinorium (12), Dashkesan Synclinorium (13), Geigel Anticlinorium (14), Agdjakend Synclinorium (15), Murovdag Anticlinorium (16), Elbekdash Downfold (17), Agdam Anticlinorium (18), Mardakert Synclinorium (19), Martuni Synclinorium (21), Karabakh Anticlinorium (22)
	Sevan-Akeri (SA)	Tourogachai Synclinorium (20), Sarybaba Synclinorium (23), Shakhdag Synclinorium (33)
	Miskhan-Kafa (MK)	Zamzur Anticlinorium (24), Khuzabirt Synclinorium (25), Lachin Anticlinorium (26), Starotag Anticlinorium (27), Gocha Synclinorium (28), Kafa Anticlinorium (29), Kelbadjar Superimposed Downfold (KSD)
	Arax (A)	Zangezur Anticlinorium (31), Ordubad Synclinorium (32), Sharur-Djulfa Anticlinorium (Sh), Nakhchyvan Superimposed Downfold (NSD), Lower Arax Lateral Downfold (AD)
	Talysh (T)	Talysh Anticlinorium

Figure 2. Chart of Tectonic Zoning

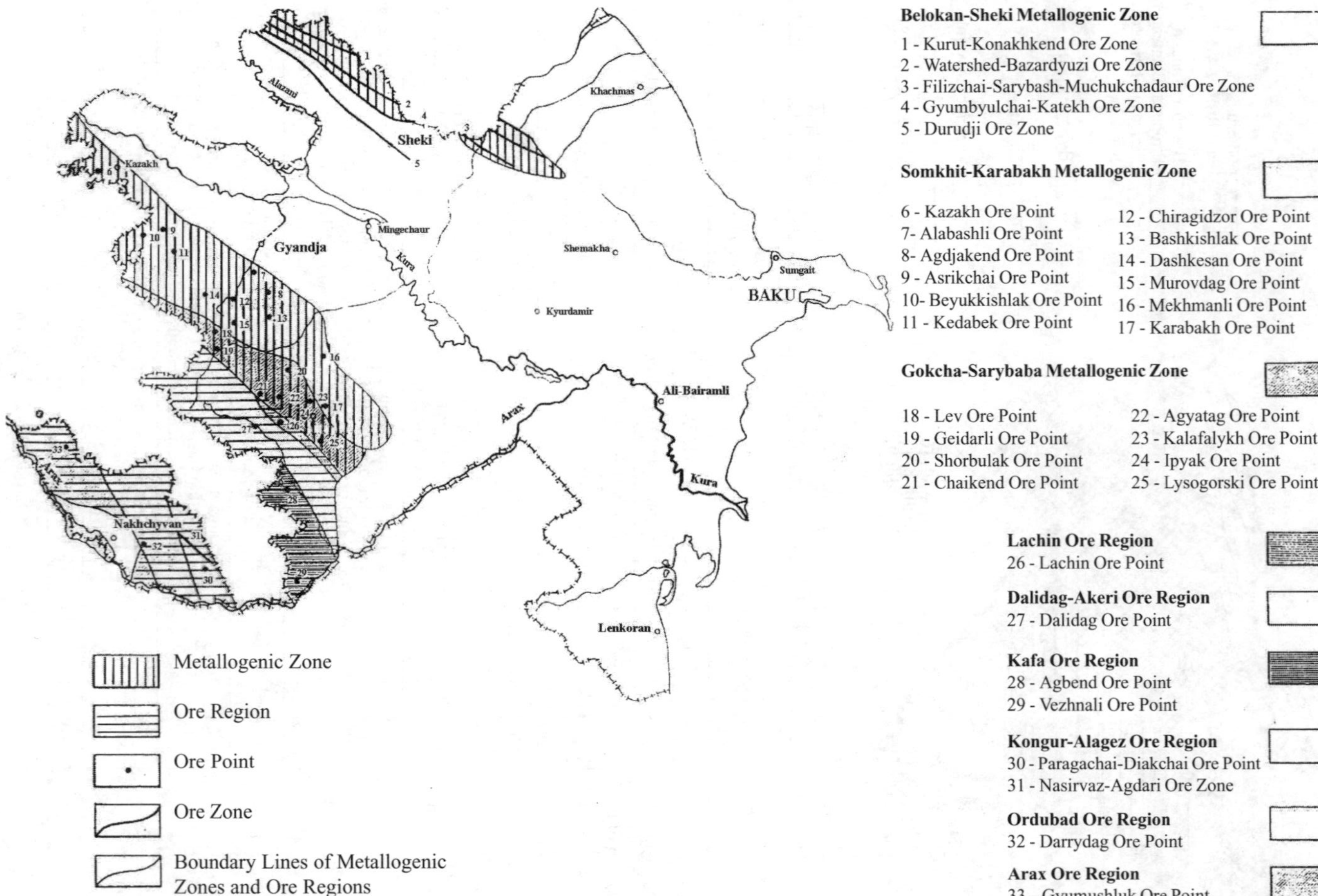

Figure 3. Chart of Metallogenic Zoning

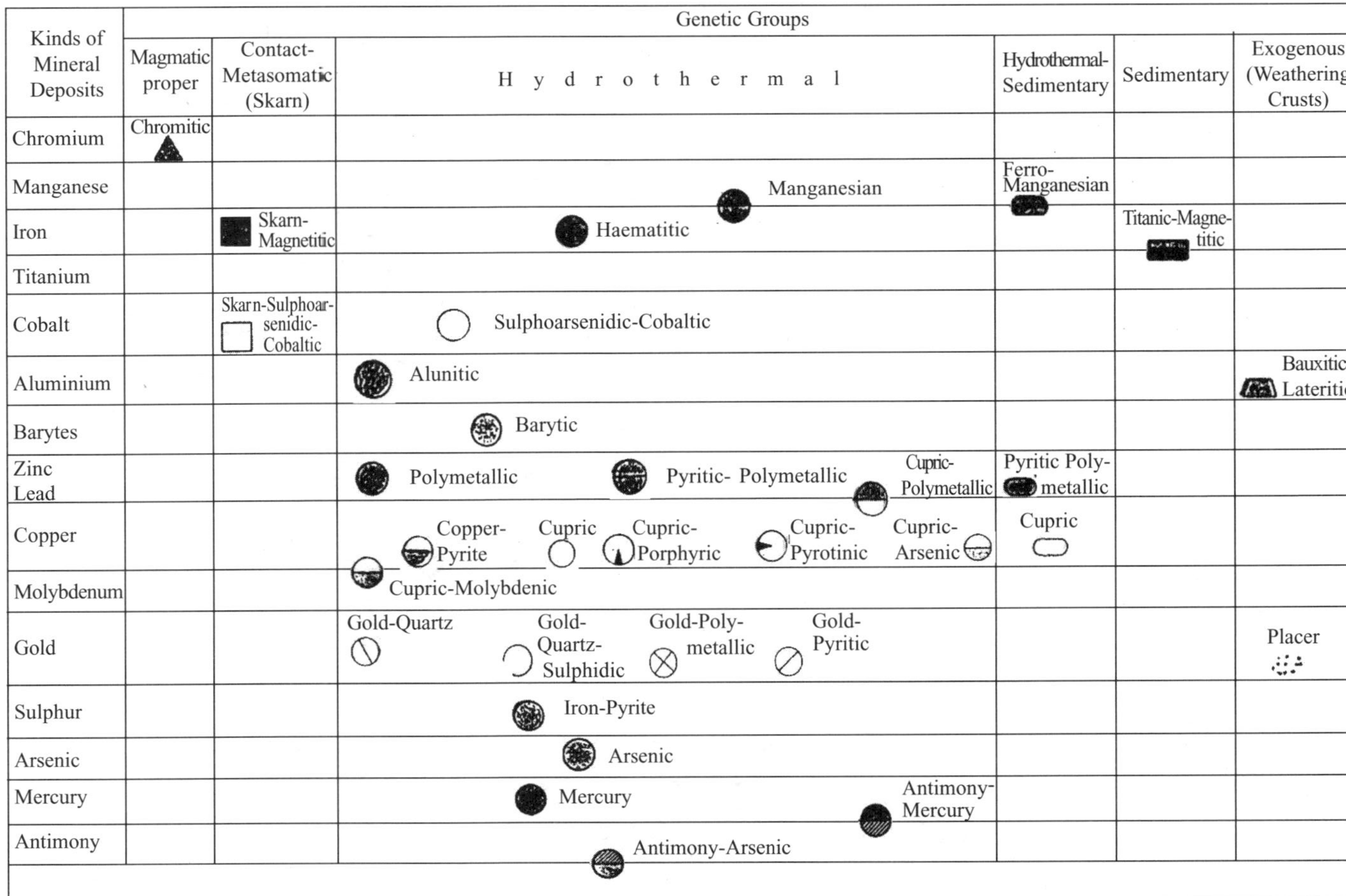

Figure 4. Kinds of Mineral Deposits, Genetic Groups and Ore Formations

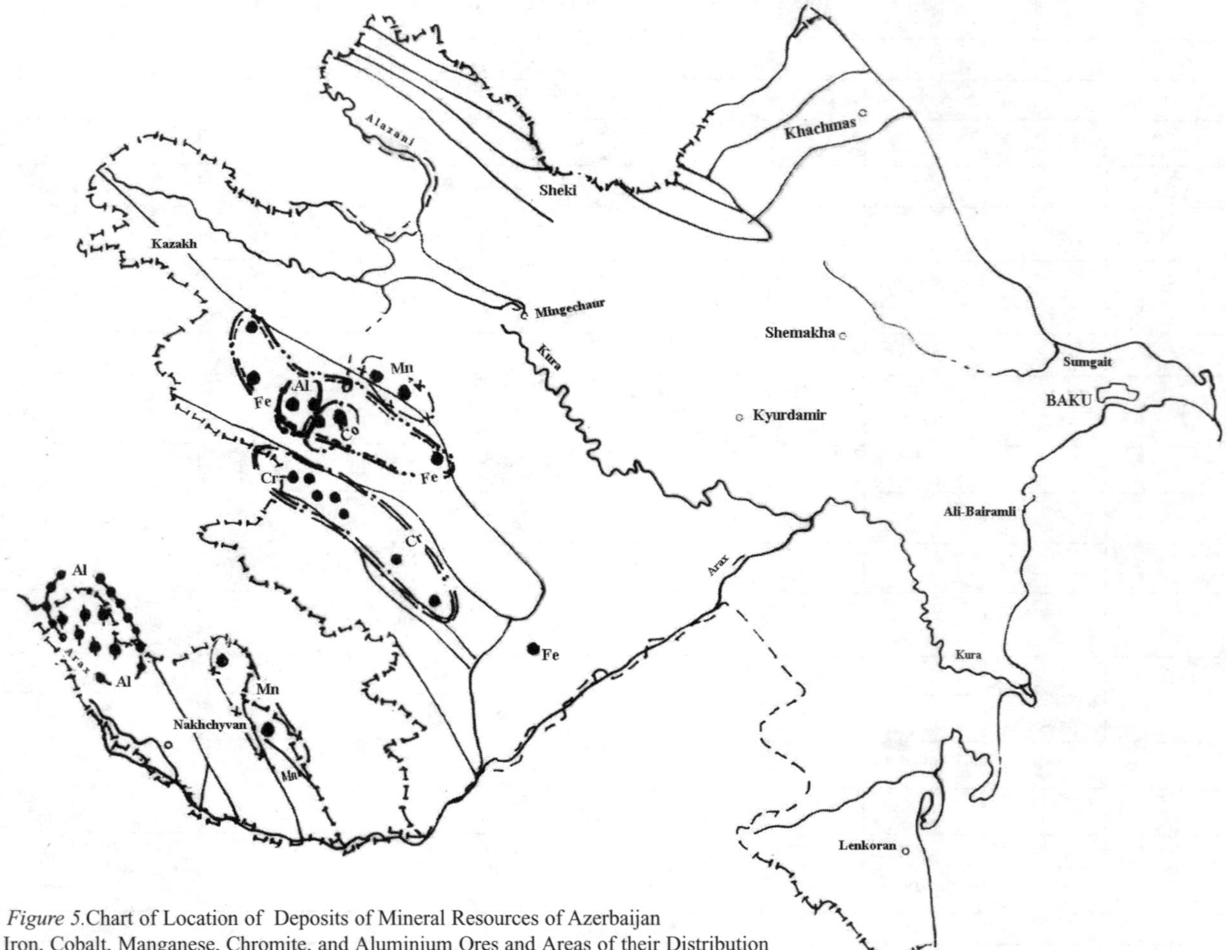

*Figure 5.*Chart of Location of Deposits of Mineral Resources of Azerbaijan
Deposits of Iron, Cobalt, Manganese, Chromite, and Aluminium Ores and Areas of their Distribution

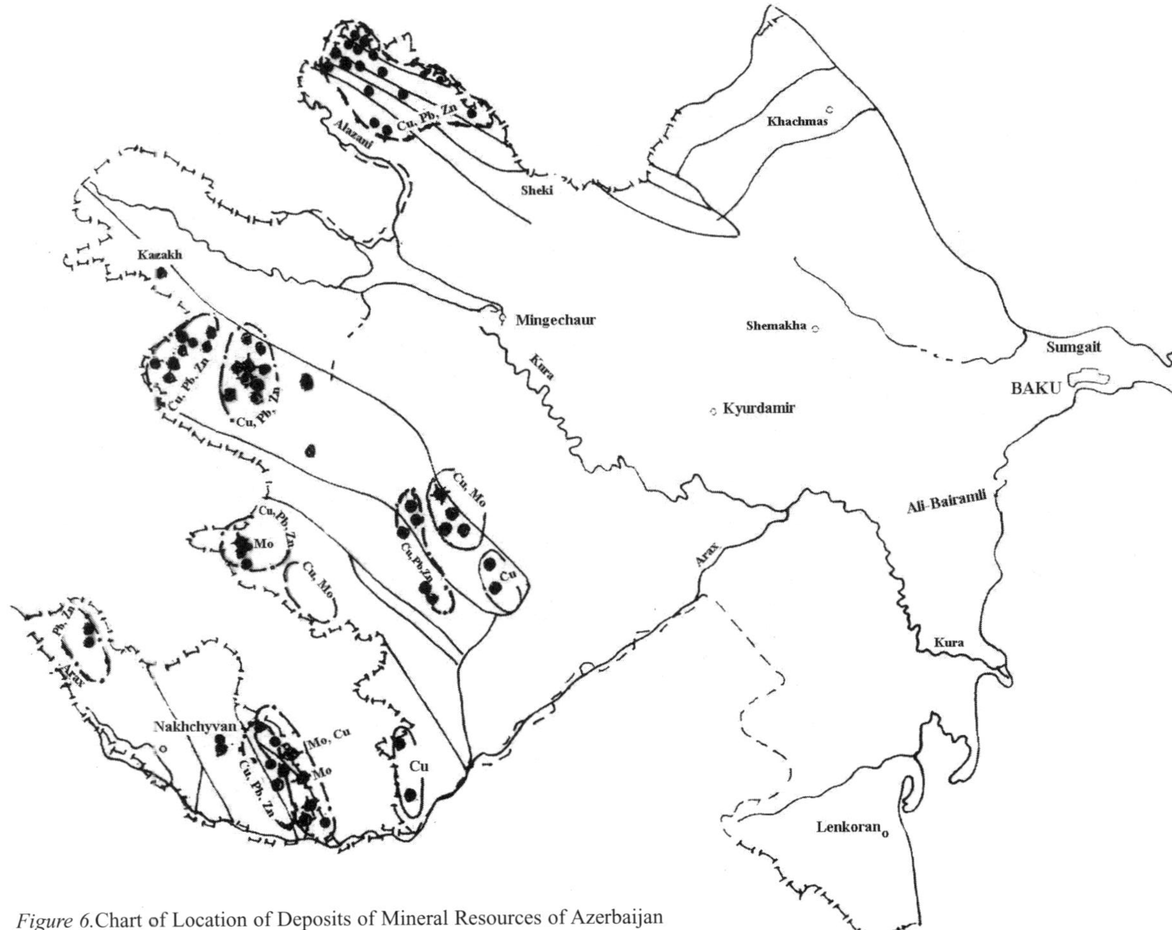

*Figure 6.*Chart of Location of Deposits of Mineral Resources of Azerbaijan
Deposits of Copper, Copper-Molybdenum, Lead-Zinc, and Copper-Lead-Zinc Ores and Areas of their Distribution

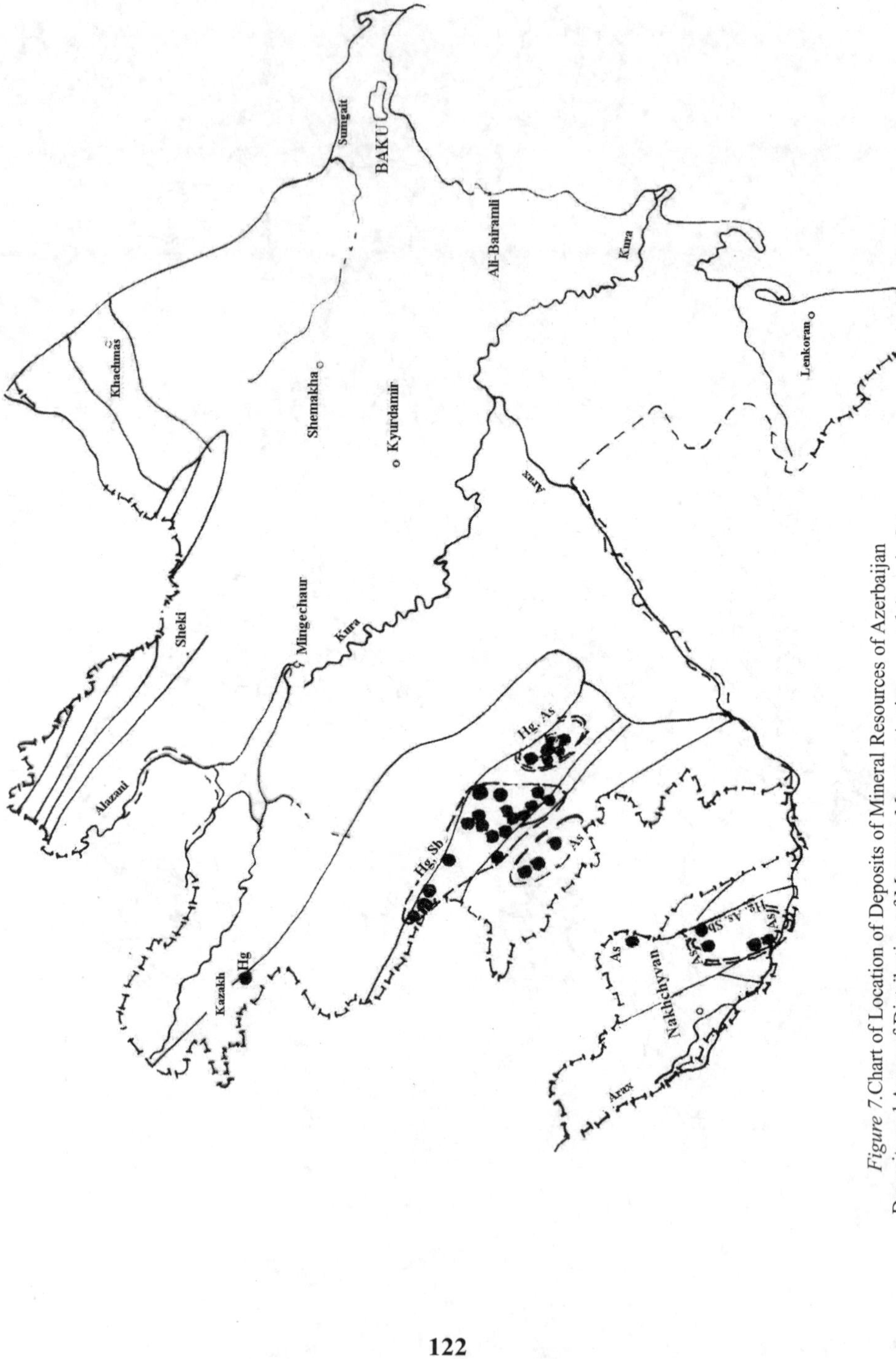

Figure 7. Chart of Location of Deposits of Mineral Resources of Azerbaijan Deposits and Areas of Distribution of Mercury, Mercury-Antimony, and Arsenic Ores

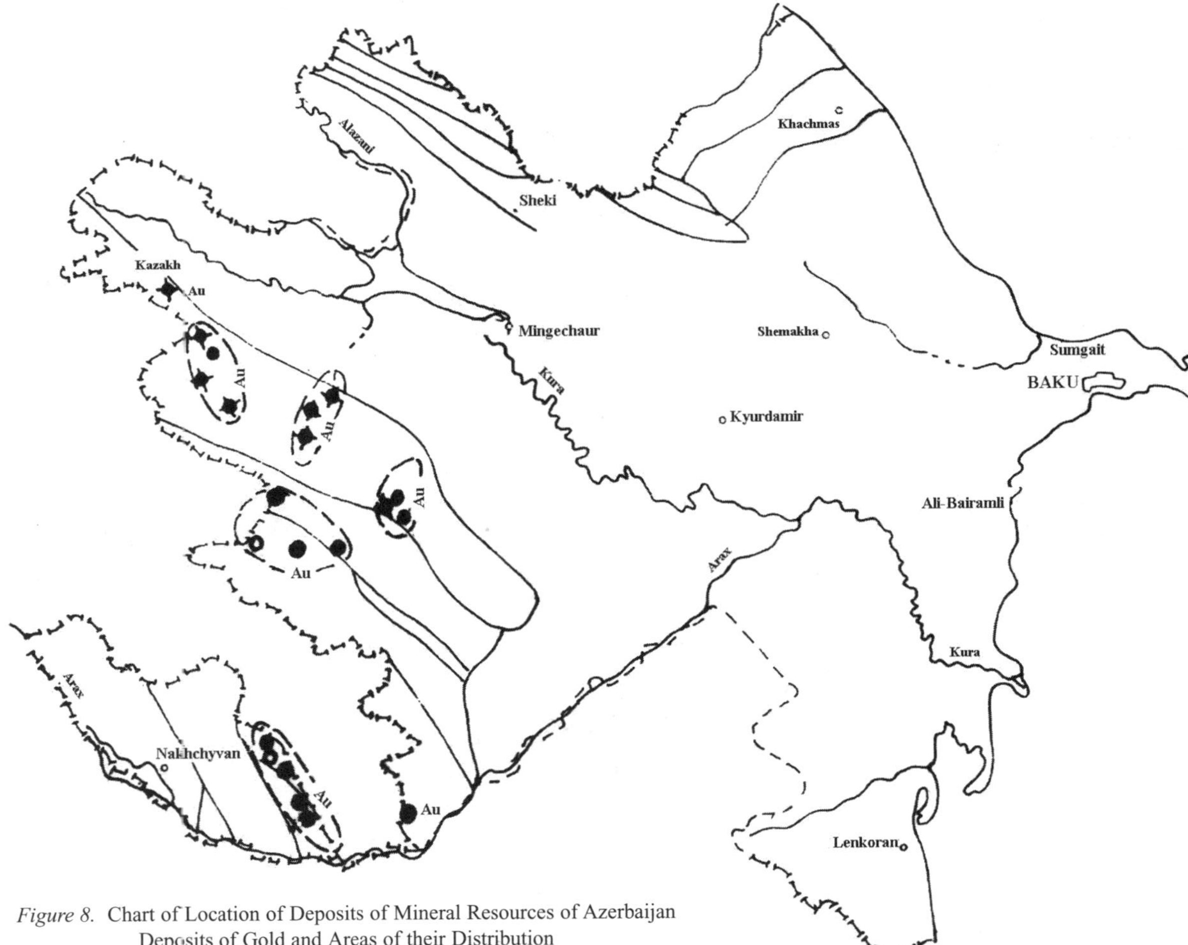

Figure 8. Chart of Location of Deposits of Mineral Resources of Azerbaijan Deposits of Gold and Areas of their Distribution

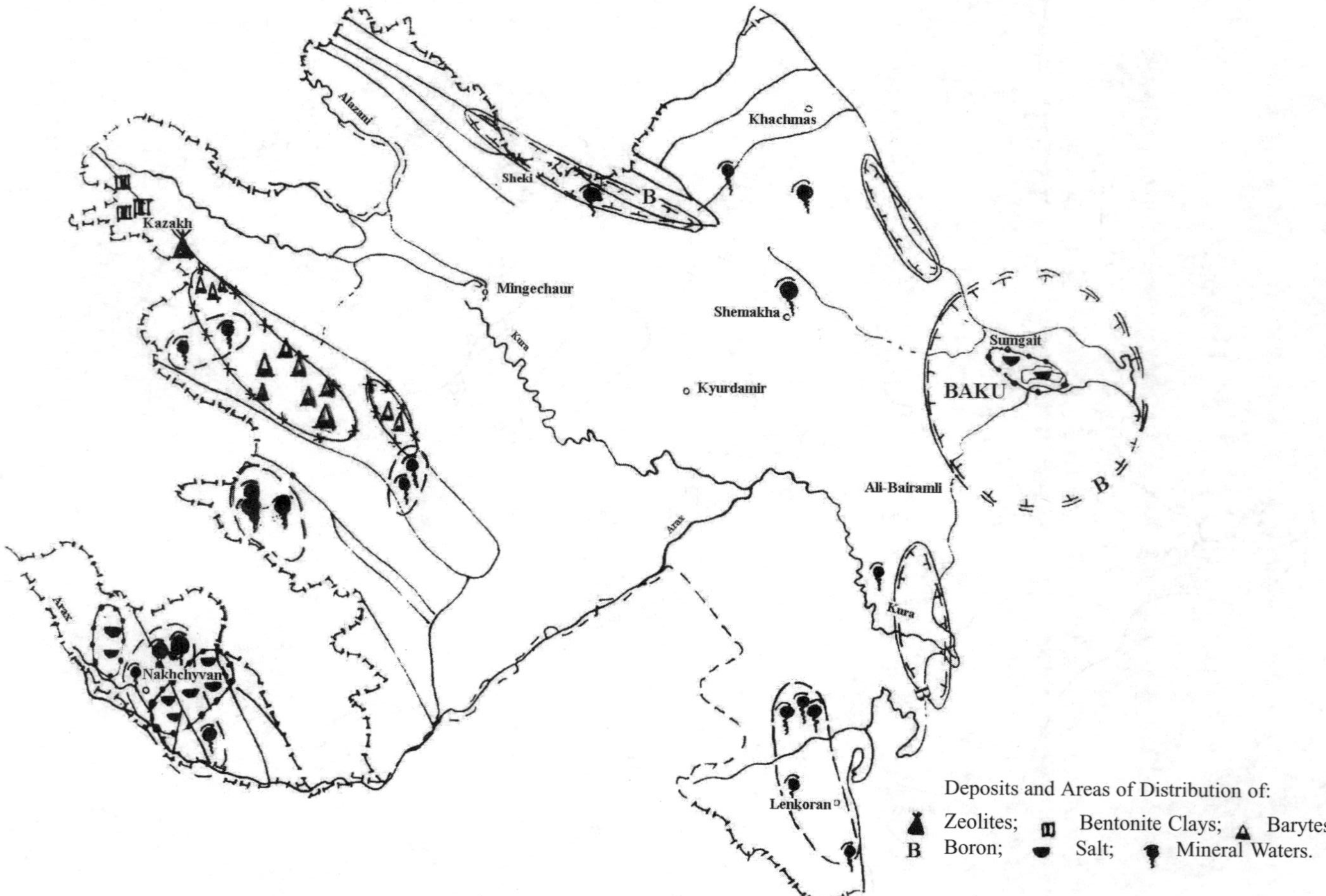

Figure 9. Chart of Location of Deposits of Mineral Resources of Azerbaijan

SUSTAINABLE DEVELOPMENT AND SYSTEMS FOR MANAGEMENT OF MINING SECTOR IN ALBANIA

G. DEMI, G. MYFTIU, G. THEMELI and J. TESTARD
Mining and Proc. Tech. Inst. Tirana,
ALBINFO, Paris France,
BRGM France

1. Geological premises and mineral potential & production

1.1. GEOLOGICAL PREMISES

Generally speaking, the structure of Albanides is divided in two large structural regions — inner Albanides and outer Albanides. Each of these regions manifests peculiar geological, tectonic and metallogenic characteristics. In the geological structuring of Albanides participates the old sediment of the Palaeozoic with a limited spreading, developing mainly in the eastern parts. The sediment of Mesozoic, mainly composed of carbonates, spreading to the south-western, central, and northern parts of Albania, has a greater development.
One of the most evident features of the geological structure of the Albanides is the spreading of the ophiolitic formation. They occupy an area of about 4300 km^2 and are represented by the plutonic and volcanogenic sequences of ultrabasic to intermediate and intermediate-acid composition. Geological and gravimetric data reveal a considerable thickness of the Albanian ophiolites, from 1-2 km in the western part up to 14 km in the north-eastern region. The most significant metallic ore deposits are linked with the ophiolite formation.

1.2. MINERAL RESOURCES POTENTIAL OF ALBANIA

The geological structure of the Albanides has conditioned a wide range and diversified natural resources deposited on the ground surface and underground. In the inner tectonic zones there is encountered the principal concentration of metallic deposits evaluated from tens to hundreds of millions tons of each deposit, viz. such as chrome, iron-nickel, nickel-silicates, coal, copper, etc.

A.G. Tvalchrelidze and G. Morizot (eds.),
Mineral Resource Base of the Southern Caucasus and Systems for its Management in the XXI Century, 125–140.

TABLE I.

No	*Description*	*Geological Reserves*	
		in million tons	Grade (%)
1	Chromite-Mineral	19.6 5.6	Cr_2O_3 18-36% Cr_2O_3 42-44%(high grade)
2	Copper - Mineral	44.9	Cu 1.3 – 1.5 %
3	Iron - Nickel	220	Fe-42.3%; Ni-0.98%; SiO_2-12.8%; Co-0.063%
4	Nickel - Silicate	102	Fe-18.3%; Ni-1.13%; SiO_2-36.9%; Co-0.043%
5	Bauxites	17	(41-52)% Al_2O_3
6	Titanomagnetites	140	5.9% TiO_2, 17.8% Fe_2O_3
7	Quartz Sands	25	
8	Gypsum alabaster	12	(88-98)% $CaSO_4$*H_2O
9	Stone salt	300	(76-82)% NaCl
10	Coal	712	3 200 kcal/kg
11	Peat	156	2 900 kcal/kg
12	Bituminous Coal	4	4 600 kcal/kg
13	Bitumen - Natural	1.4	
14	Bituminous Sands	247	8-10% bitumen
15	Olivine	59	MgO-48%; SiO_2-37%
16	Dolomite	40.6	CaO-30-34% , MgO-18 — 215
17	Limestone	Unlimited	
18	Marble & Dimension Stone	130	CaO - 50%;
19	Basalt rocks	Unlimited	SiO_2 - 47 %; Al_2O_3-13

1.3. MINERAL PRODUCTION

The industrial exploitation of the solid minerals in Albania began before the Second World War in the form of concessions given mainly to Italian foreign companies. Mineral Industry after the World War II was established as a State Enterprise based on the exclusive rights of the State over the explored and unexplored mineral deposits and was developed on the basis of a total exploitation of the mineral resources ignoring the criteria of economic efficiency, protection of the environment, etc. To this end Albania was developed as an intensive mining country consisting of numerous mines and processing units of chrome, copper, coal, ferronickel, limestone, bitumen, tar sand, etc. During the period between 1986-1996 Albania produced about 7.1 million tons of chrome ore, 6.8 million tons of copper ore, 5 million tons of iron-nickel ore, 12.5 million tons of coal, 1 million tons of bituminous coal, 100 thousand tons of natural bitumen, 500 thousand tons of bituminous sands, 250 thousand tons of olivinite, and more than 25 million tons of limestone.

After 1991, the transition period caused a sharp decline of Albanian mineral production caused by:

- Gradual reduction of geological reserves of industrial importance.
- Transition period itself and the confrontation of the Albanian mining production with world market competition.
- Poor management at enterprising and ministerial levels.

TABLE II. Mineral production of the last five years

			1996	*1997*	*1998*	*1999*	*2000*
1	Lumpy Chrome Ore	000/ton	236.358	157.203	150.285	79.445	61.087
2	Copper Ore	000/ton	187.765	24.815	53.477	33.945	0
3	Nickel Silicate	000/ton	0	0	0	0	12.000
4	Bauxite	000/Ton	0	4.454	4.128	4.624	1.470
5	HC Ferrochrome	000/Ton	31. 00	34.00	30.00	28.20	25.00

TABLE III. Production Forecast For Future Years

			2001	*2002*	*2003*	*2004*	*2005*
1	Chrome Ore	000/ton	80.0	105.0	125.0	140.0	140.0
2	Copper	000/ton	100.0	240.0	240.0		
3	Nickel Silicate Ore	000/ton	50.0	100.0	300.0	300.0	300.0
4	Bauxite	000/ton	5.0	10.0	20.0	30.0	30.0

2. Legal Framework

2.1. LEGISLATION OF MINING SECTOR

Different legal actions were carried out starting from 1993 toward restructuring and privatization of the mining industry. The most important ones are the following:

- The approval of the Mining Law of Albania No. 7491, on 17.02.1994 accompanied later on with the respective Acts and Regulations for licences and activities of prospecting, exploration and exploitation.
- Resolution of the Ministers' Council on 21.03.1994 on the approval of implementing Discount Cash Flow procedures Method for Evaluation of the Mining Assets for the purpose of privatization of mining industry (approved by the Parliament as well).
- Law No 8026, date 9.11.1995 "On the Privatization of the Commercial Societies, which operate in the Mining Sector".
- Law No. 8306, date 14.03.1998 for Privatization Strategy of particular strategic sectors (including mining sector).
- Council of Ministers Decision (VKM) No 421, date 09.07.1998 and the Order of Minister of Public Economy and Privatization No 9, date 01.09.1998 *"On the treatment of exploited mining deposits as well as of those mines which will be separated from the commercial societies and state enterprises"* and the document of the General Directory of Mines No 4631, Prot. date 04.09.1998 Applying the Minister's Order No 9, date 01.09.1998.
- The law on the concession of the mining enterprise to foreign and local entrepreneurs 1998-1999.
- Law on Collateral approved recently to help banks and mining SME finance projects.
- Law No 8761, date 02.04.2001 for Approval of concession agreement of "BOT" form between Minister of Public Economy and Privatization and Turkish Company "BER-ONER" for some Copper and Chromium Industry Objects and for the giving of some incentives and guaranties to concessionaire of this agreement.
- Law No 8791, date 10.05.2001 for Approval of concession agreement of "ROT" form between Minister of Public Economy and Privatization and Italian Company "DARFO" for Bulqiza Chromium Mine, Bulqiza Chromium Dressing Plant, Klos Chromium Selection Impiant and Burreli Ferrochromium Metallurgical Plant and for the giving of some incentives and guaranties to concessionaire of this agreement.

TABLE IV. Mining taxation
Data Source: Ministry of Public Economy and Privatization, Directory of Statistics

	Taxes in %
1	Custom taxes: Import of machinery 2%. Import of goods 15%. Export production - no customs tax
2	VAT VAT tax based on law No 7928, date 27.4.1995 is paid 20% for all activities and is reimbursed. **no VAT tax for export goods**
3	Profit tax for all activities 25%
4	Royalty based on the mining law No 7799, date 17.2.1994 art. 83 is 2% of value of sales of minerals...

2.2. INSTITUTIONS IN MINING SECTOR

Under the authority of MINISTRY OF PUBLIC ECONOMY AND PRIVATIZATION there are:

- The General Directory of Mines.
- The Directory of Privatization.
- The Directory of Negotiations.
- Juridical Directory.
- Mining and Processing Technology Institute (Technical Economical Consulting and Negotiating Staff).
- Albanian Geological Survey.
- Geo-Technical Consulting Staff.

The co-operation between each of the above institutions and directories has made possible the preparation for privatization process so far and could ensure the follow-up and continuation of the process.

2.3. ALBANIAN KNOW-HOW IN MINING INDUSTRY

The Discount Cash Flow evaluation for mining project has been approved by Council of Ministers Decision No 115, date 21.03.1994.
This method is well-known in the world and is widely used, being compatible with international standards for negotiations on a common Know -How base.
However, the evaluation based on the traditional inventory is still used for partial privatization of mining assets, buildings or machinery.
Albania is a known place to find skilled miners and engineers especially for small mines of chrome deposits with special geological features and structures. The mining staff has the experience to work under difficult conditions and is able to cope with any geological and mining environment. The production records show that for almost half a century intensive mining has been the main culture work and tradition in many mining areas.

3. Privatization of Mining Industry of Albania

3.1. GENERAL CONSIDERATIONS

A comprehensive privatization programme is a key to Albania's progress towards a developed market economy. The first law on privatization was approved in August 1991. The law allows a wide range of methods including auctions, concession bids, direct sales, free distribution of shares, etc.

Small companies SME-s have been disposed of largely through employee buy-out or auction. Nearly all small-medium enterprises (SMEs) that had not been privatized during the early years of the process were sold or liquidated during 1997 and 1998. Many of them were economically not viable, hence they were liquidated and their assets were transferred to local authorities.

In March 1998, a more flexible regulatory framework for privatization was approved and in May 1998 a draft privatization plan was released, according to which the publicly owned businesses could be sold below book value, which, in many cases, does not reflect the market value. This will enable the Government to divest the state share in joint ventures to private sector counterparts and private enterprises in strategic sectors. The Government intends to privatize (at least partially) all main state monopolies, including transport, telecommunications, energy, mining and water. Currently, the Government is looking for strategic investors for these sectors. The privatization method will be by international tender on a case-by-case basis. The cash received will be used to reduce the budget deficit.

Some of the big companies included in the privatization list are:

- Telekomi Shqiptar - Albania's fixed network telephone operator.
- Albanian Mobile Communications (AMC) - The only mobile phone operator in the country (already privatized).
- Korporata Energjetike Shqiptare (KESH) - The state electricity producer.
- Albpetrol - The state oil company, which currently has joint venture agreements with several foreign oil companies.
- Albkromi - The state chrome mining company (completely privatized to DARFO, Italy and BER ONER, Turkey in the concession form).
- Albbaker - The state copper mining company (already privatized in the form of concession to Turkish Co. BER ONER).
- Savings Bank of Albania - The second biggest bank in Albania, etc.
- National Commercial Bank of Albania - The second biggest second-tier bank in Albania (already privatized).

The impossibility for the State administration to successfully manage (in Albania as well as in any other country of the world) the mineral potential imposes the necessity to privatize the mining industry.

The efforts to rehabilitate the Mineral Industry started in 1993, resulting only in closing

the majority of coal and ferronickel mines and some of the smaller copper mines, etc. The Albanian State (Ministry of Public Economy and Privatization) is still the main owner of the former production mines and processing units in chrome and copper industry, established in the form of ALBKROM sh.a. (State-owned shareholding chromium company) and ALBAKER *(Albanian Copper Corporation)*.
These two State companies, ALBKROM and ALBAKER proved to be non-effective production enterprises after 1990 and faced many problems linked mainly with the reduction of labour force causing negative social phenomena, such as unemployment.
The privatization process, proposed to be carried out in Albania beginning from June 1995, is just completed in many other countries and is underway in some other countries of the Central and East Europe. Apart from the duration in time of the mining industry privatization process, it seems to be an irreversible process and a major trend, as it has produced positive results, where it has been implemented.
The end of the year 2000 marked the start of the revitalization of chrome industry through the birth of many small chrome mines which brought the most skilled miners and engineers in the mining areas back to employment. But the process of mining growth in Albanian chrome industry is still moving slowly because of the limited financial possibilities of the small chrome mines and the scarce information the Albanian private banks have on chrome and mining industry potential of Albania. This project will also help to inform banks on the current progressive growth of the private mining SME in chrome and other sectors, such as construction materials.

3.2. PRIVATIZATION TARGETS

Privatization targets in Albania are summarized as follows:

- Maximum attraction for the initial investments from potential local or foreign investors, without ignoring the continuation of the existing investments.
- The presence of big strategic mining companies, which are expected to remain for a long time and to apply modern technology, without damaging the environment.
- Increase of the stream revenues from private mining sector, in combined forms of taxation, royalties, custom duties, etc.

The main target might be focused on re-vitalization of the mining industry through privatization process in order to increase the total profitability and the revenues for the people employed in mining industry.

3.3. PRIVATIZATION PROCEDURES IN MINING SECTOR

The privatization procedures could be envisaged mainly as preparatory and executive ones.
The preparatory phase:

- The division of the big mining units into a number of units of different size and separation into independent ones.
- The drafting out of the rules of privatization.

- The drafting out and selection of the type of contracts for privatization.

The executive phase:

- The call for the bidders to buy or reward the license of exploitation.
- Establishment of private mining operators.

The division of the big state-owned mining units (ALBKROM) and separation into many independent ones is carried out on the following criteria :

- *Geographical-regional criterion:* Different geographical regions will define different objects for privatization.
- *Geological criterion:* Different deposits will define different objects offered for privatization.
- *Environmental, safety & technical-mining criterion:* This criterion is important to avoid incidents and increase industrial reserves by exercising strict rules and agreements between private mining companies working close to one another.
- *Organizing criterion:* It is intended that any private mining company should be as independent from the others as possible. This could be achieved for example by avoiding the use of the same transport routes.
- *Social criterion:* It is intended to provide the conditions for the creation of as a great number of private mining companies as possible, aiming at the creation of the maximum possible employment.

The above-mentioned criteria applied to the division and separation of mining industry resulted in the creation of the small and large mining units. It is recommended that a sustainable and suitable option of privatization should correspond to each class of mining unit.

3.4. POST-PRIVATIZATION

Attention in this programme of promoting private mining SME is drawn also to the ***sustainability and environmental*** exploitation of the mineral resources. That means, that in order to ensure a normal running of the licensed private mining companies **the State will retain control of** the following issues related to the private mining activity:

- Completion of the missing geological and other mining data.
- Follow-up of existing environmental regulations and the new ones to be up-dated.
- Harmonization of Albanian regulations with EU Mining Directives toward short term and long terms compatibility

Other issue, accompanying the birth of private mining companies is the **local authorities**. Albania is a country, where local authorities are still incapable to run the development of natural resources, especially major and specific projects like those of mining, oil, energy, water, etc. The co-operation between Institutions, the State (General Directorate of Mines) and consultants is of critical importance for the operation of private mining SME.
The division of the ALBKROM SH.A. created a limited number of big chrome mines and numerous opportunities for private mining SME.
Generally there was not more than one application for each small mine because the separated small mines were already exploited many years ago. But some of them

represent interesting profitable mines for a mining SME. In such cases all applications are subjected to a tender /competition. Advantages are given to the companies represented by former mining staff. Anyway, the wisest decision as to whom to give the mine should be taken in co-operation of the local authorities with the national authorities (Ministries) & national research centres and Institutes.
The best ways of privatization are considered to be the cases of a combination of international investors and local experts and labour employment. DARFO – an Italian Co. and BER ONER – a Turkish Co. have so far succeeded in this respect.
The big chrome or copper mines are subject to negotiations based on the Concession Law and Mining Law.
Since there is no intense competition so far, the negotiations start once it is proved that the applying company has the capacity to run and finance the project. In cases when more than one company apply for a mining concession or privatization, then, a bidding procedure is organized under the supervision of the Ministry of Public Economy and Privatization.
Recent negotiations have finalized the Copper Industry Concession to BER ONER – Turkish Co., while the Chromium Industry of Bulqiza Chrome mine (4 million tons of industrial reserves) and Burrel Fe-Cr smelter are transferred to DARFO Co. In the latter case two Italian companies (DARFO and METAL LEGA) were in competition. Another concession agreement is under negotiation process for ***the areas of nickel deposits*** in the south-east of Albania with the Australian company Adriatic Nickel.
Another issue of Privatization is ***further unemployment in the first phase***. The Trade Unions do not agree with it. That is why there must be found a specific strategy to negotiate with the leadership of Trade Unions in order to convince them that the privatization cannot be stopped and that it is absolutely indispensable for its positive impact on economy.
The experience has shown that the Trade Unions may become the major and most conplicated factor, which can stop the privatization process for some time, as it has happened in Germany and Romania recently. In any case, it is necessary to find a sustainable formula of privatization associated with social assistance to the unemployed working force. A ***special law*** was approved by the Government of Albania to the effect that miners will be paid from one to two years if they are fired during the privatization or mining closure procedures.
Recent gradual growth of private mining industry in the sector of chrome and construction materials had the positive impact on employment of the most skilled workers and engineers.

4. Private Investments in Mining Industry of Albania

4.1. FOREIGN INVESTORS IN ALBANIAN MINING INDUSTRY DURING 1990-2000

The above-mentioned actions taken mainly toward the legislation improvement and privatization reform aroused the interest of the following ***foreign companies*** willing to have a venture in Albania (see in table V.):

TABLE V.

Name of company	*Country*	*Year*
UNION CARBIDE	U.S.A.	1992
ILVA-S.P.A./TECHINT	Italy	1992
I.T.S. GROUP (ACCIAIERIE VENETE)	Italy	1992
CONSIDAR EUROPA	Belgium	1992
A.W.T.	Austria	1992
MARC RICH + CO A.G.	United Kingdom	1992
SUMITOMO CORPORATION	Japan	1992
B.S.E.-ESSEN G.M.B.H.	Germany	1992
MACALLOY CORPORATION	United Kingdom	1993
FUCINATI S.P.A.	Italy	1993
PREUSSAG	Germany	1995-96
ACCIAIERIE VENETE	Italy	1995-96
KINGLOR S.R.L.	Italy	1997-1998
ROBINCO S.A.,	United Kingdom	1997-1998
NEBEX	Canada	1995-1998
KARMA ROYAL ROADS	Canada	1996-1998
AQUATER & DESCAT	Italy	1996-1998
TRIAKOM	Germany	1997
HAYRI OGELMAN MADENCILIK	Turkey	1995-1999
LARCO.S.A.	Greece	1997-2001
DARFO	Italy	1994-2000
TESARO	Italy	2000
ADRIATIC NICKEL	Australia	1999-2000
BER-ONER	Turkey	2000
METAL LEGA	Italy	2000

Some of these companies, such as NEBEX, AQUATER, KARMA ROYAL ROADS, DARFO, ADRIATIC NICKEL, etc. have concluded the negotiations by being granted licences to practise prospecting, exploration or exploitation.

NEBEX RESOURCES LTD, a Canadian company listed in Toronto Stock Exchange concluded a deal with ALBAKER (Albanian Copper Corporation) in December 1997 for Munella copper/gold/polymetallic deposit. The deal mainly refers to the transfer of the licence of exploitation from ALBAKER to NEBEX for a payment of 2.5 million USD success fee to ALBAKER on developing Munella gold/copper/polymetallic deposit. In 1998 NEBEX RESOURCES LTD went bankrupt in Toronto stock exchange, cancelling the programme after paying $ 2.2 million.

The failure of the majority of foreign investors listed above to achieve a deal in Albania during 1992-2000 is due to the following reasons:

- **In spite of the good legislation approved in 1994 and 1995, the governmental staff and the majority of local experts were not convinced of the benefits of privatization.**
- **Lack of negotiating skills on the part of the governmental staff.**
- **Lack of political stability during 1997-1999 (including the Kosovo crisis).**
- **Objections on the part of the Trade Unions against privatization due to immediate unemployment effects.**

Anyway the year 2000, with a general post-conflict stability (and Balkan Stability Pact) brought about a gradual mining growth and some positive developments in mining industry with regard to foreign and local investments like the following:

Current /recent growth of foreign investments in mining industry -

DARFO Italy got the chrome concession (Elbasan Ferrochrome smelter and Katjel chrome mine). This concession based on a 15 million USD investment project is the only actual success, connected with foreign investors so far.

LARCO.S.A. **Greek Company has concluded a deal on Bitincka nickel deposit to carry out an industrial test of 50 000 tons.**

Adriatic Nickel**, a company of Australian origin is still in negotiations for main chrome deposit of Bulqiza and nickel silicate deposit of Bitincka.**

Another *French company* is negotiating for bitumen and tar sand deposit of Selenica near Vlora (the Southwest of Albania).

Recently DARFO and another Italian group METAL LEGA ***were bidding for the concessions*** of Bulqiza chrome mine (industrial reserves of 4 million tons of rich chrome ore, the biggest European chrome mine and one of the biggest chrome deposits of the world) and Burrel Fe-Cr smelter. The bid evaluation named DARFO Co as the winner of the competition.

Current /recent Local Albnanian Investors in mining industry -

Apart from what has been said about the above-mentioned foreign investors, on the basis of the Mining Law of Albania (1994), during the period between 1994-1998 there started the establishment in the country of *private mineral production sector*, as local *Mining SME* started working mainly in limestone quarries for construction materials, chrome quarries and mines, tar sand, bitumen, marble, dimension stone, etc.

The privatization process of the former state-owned mining enterprises (ALBKROM), especially in the last three years (1998-2001), brought in a new generation of private mining entrepreneurs.
There are about 300 Albanian private mines operating mainly in the field of the production of construction materials and chrome ore, as well as that of trading.

4.2. OPPORTUNITIES AND INCENTIVES FOR FOREIGN INVESTORS

Notwithstanding the difficulties and challenges, Albania is a country that offers opportunities to foreign investors.
The already mentioned privatization process offers a wide range of opportunities. Potentially high-profit sectors are mining and oil extraction. Albania is the only country in Europe possessing substantial reserves of chromium. Before 1990 the country was the world's third largest producer of chrome ore. Both oil and mining industries have been export-oriented. However, significant investments are needed in order to modernize the production methods. Other areas of interest include transport and telecommunications, agriculture and food processing, textiles and clothing, and hydropower.
Foreign investments in Albania are fully protected by law. They cannot be nationalized, expropriated or subject to any other measure, except for special cases provided by law, such as public interest, etc.
Moreover, to help Albania provide financial security for private sector entities investing in the country, the World Bank has established a Political Risk Guarantee Facility (PRGF). The Albanian Guarantee Agency (AGA) administers the PRGF and enterprises involved in productive activity are eligible for applying for guarantees (excluding tobacco products, alcohol and armaments).
Specific categories of risk covered by the Political Risk Guarantee Facility are:

- Inability to convert and transfer currency.
- Expropriation.
- Seizure of goods, prevention of sales or prevention of exports/imports.
- War or civil disturbances.
- Cancellation of licenses and restrictions on import and export.
- Imposition or increases of import or export taxes is as a consequence of war and civil unrest.
- Interference in the carriage of goods.
- Diversion of voyage.

Other legal incentives include:

- Equal treatment of foreign and domestic investors.
- Full profit and dividend repatriation (after taxation).
- Funds from the liquidation of a company may be repatriated.
- Albania has signed a number of bilateral agreements on the promotion and protection of reciprocal investments with Austria, Bulgaria, China, Croatia, Egypt, France, Germany, Greece, Israel, Italy, the Netherlands, Poland, Romania, Sweden, Switzerland, Tunisia, Malaysia, Turkey, the US and the UK.

- Double taxation is in force with the following countries: Poland, Romania, Malaysia, Hungary, Turkey, Czech Republic, Macedonia, Russia, Croatia, Italy, Bulgaria and Norway. There are treaties signed, but not yet in force with Greece and Switzerland. With Egypt, Belgium and Malta the double taxation treaties are initialled on technical level.

The new Albanian system of secured financing on movable property for Albania came into effect on February 1, 2001.

The new *Law on Securing Charges* is designed to generate the needed credit so that to allow the new market economy in Albania to develop. This is accomplished through a mechanism that reduces the risk involved in advancing credit by providing security using the **debtor movable property**. Credit grantors are more likely to advance credit as risk decreases, while borrowers will be able to use such credit to increase their inventory and equipment in case of business, or to purchase new consumer goods in the case of consumers. It is expected that the agricultural sector will be one of the main beneficiaries of the new system.

The new Albanian Registry of Securing Charges provides public disclosure of existing interest in movable property including claims resulting from security agreements, lease agreements, conditional sale agreements and transactions with accounts. Such disclosure will protect persons wishing to acquire a property right in the debtor's property. Public disclosure of such property rights through registration will notify potential creditors or purchasers of property of the existence of earlier security interest in it. A person, who will desire to acquire a property right in the property, will have the means to discover whether or not it is already encumbered by a property right. The new Albanian computerized registry system also provides fast, easy and inexpensive access, thereby guaranteeing that registration and searching costs will have little or no effect on the decision of parties to engage in a secured financing arrangement. Finally, the new Albanian secured financing system on movable property recognizes the time and costs involved in recovering on defaulted agreements and, accordingly, provides expeditious and inexpensive methods for enforcing security agreements through seizing and disposing the debtor's property.

It may be expected that more credit will be available for mining, businesses as well as to other borrowers in Albania and a steady growth of the Albanian economy will follow.

In the end of January, 2001 in Tirana there was opened an office of registration of mobile assets (i.e. vehicles, machinery) which could help local private investors to increase the asset warranty required by the banks.

5. Market considerations

5.1. GENERAL CONSIDERATIONS

In 1997, like many key economic indicators, due to the social unrest, the foreign investments fell by 50% compared with the previous year. However, 1998 presented clear positive signals.

During this period, foreign direct investments principally supported by international financial institutions, such as the World Bank, the IMF and the EBRD, were estimated at USD 95 million.

Foreign trade increased by 20% compared to 1997. During this period, exports grew at a rate of 31% and 17%, respectively, due to the growth of exports related to mineral and alimentary products.

Foreign trade still remains oriented towards countries like Italy, Greece and Germany with some 86% of exports, whereas imports from these countries amount to 76% of the merchandise.

Italy is considered the biggest trade partner of Albania, Greece ranking second.

These two countries have also a major presence in Albania through 1,532 joint ventures that operate mainly in the service sector, and 890 wholly owned foreign companies.

About 500 Italian companies have invested over USD 100 million mainly in the clothing and shoe industry, as well as in furniture production and fishing.

The table below regroups the data of the Albanian Statistics Institute - INSTAT, relative to the pattern of foreign investments from 1993 to 1998 (number and percentage of companies):

TABLE VI.

Activity/Year	*1993*		*1994*		*1995*		*1996*		*1997*		*1998*	
	No.	%	No.	%	No.	%	No.	%	No.	%	No.	%
Agriculture	1	0.3	3	0.4	7	0.7	11	0.9	12	0.9	14	1.0
Industry	50	13.9	101	14.5	129	13.4	177	14.7	202	15.5	244	17.2
Construction	15	4.2	35	5.0	56	5.8	63	5.2	72	5.5	84	5.9
Service	31	8.6	58	8.3	94	9.8	133	11.0	151	11.6	159	11.2
Transport	19	5.3	45	6.4	51	5.3	56	4.6	60	4.6	61	4.3
Trade	243	67.7	456	65.3	623	64.9	766	63.5	808	61.9	858	60.4
Total	359	100	698	100	960	100	1206	100	1305	100	1420	100

The table below indicates the main trading partners of Albania and the export-import data for the first trimester of 1998 compared with the previous year (Source INSTAT):

TABLE VII.

Exports			*Imports*		
For the first quarter of 1998			*For the first quarter of 1998*		
In 000/USD			*In 000/USD*		
Country	Amount	% of change with previous period	Country	Amount	% of change with previous period
Italy	80.5	82.95	Italy	253	43.18
Greece	32.7	55.71	Greece	175.1	109.45
Germany	9.2	22.66	Germany	22.2	45.09
FYROM	2.2	(15.38)	Turkey	16.8	(1.75)
Slovenia	2.4	140.00	Bulgaria	14.8	39.62
Belgium	2.1	(43.24)	Slovenia	8.4	55.56
France	2	0.00	FYROM	8	17.65

5.2. CONSTRUCTION MATERIALS MARKET

The *boom of construction industry* in the western part of Albania (especially in the regions of Tirana and Durres, where more than half of population actually lives) brought about the operations of many private mining operators mainly on limestone quarries. Hundreds of crushers were imported and installed all over Albania to supply infrastructure and housing projects. Efforts are under way to establish brick factories, plants of pavement tiles, prefabricated construction elements, ready-made concrete plants, bitumen asphalt plants, etc.

The immediate increase of demand for raw materials, such as sand and crushed stone caused the exploitation of the gravel from the river basins, which is still in process, causing environmental and erosion problems. Tens of thousands of cubic metres of sand and gravel are excavated from riverbeds. *As a market and business indicator* it means that there are opportunities for limestone quarry business. *From the environmental point of view* the exploitation of the riverbeds must be stopped as soon as possible. Experts and NGO-s draw the public attention to that issue and the formation of public awareness is also in process. But before the river gravel operations will be stopped, new limestone quarries should be opened to face the rising market demand.

On the other hand, however, in order to meet the demands of the Albanian construction market, there are *imported significant quantities of tiles, pavement tiles, bricks* as well as other construction elements, although the raw materials, such as clay, limestone, olivine, etc. are on the ground.

5.3. CHROME MARKET

Albania is well known in the world as a producer and trader of chrome.
As the main solid mineral of Albania, chrome played an important role in the country's economy for more than 30 years, achieving world production peaks in 1989 with about 1 million tons, ranking the world's second largest chrome ore producer country after South Africa.
After 1990 the production of chrome had a sharp decline because of the transition and the reduction of industrial reserves. But there are still few world-class chrome deposits and many medium and small chromium deposits with a turnover potential of several hundred millions of USD.
Europe is a primary consumer of chrome ore and ferrochrome. Only Italy needs more than 200,000 tons of ferrochrome per year. The geographical position of Albania, the geological and industrial reserves offer opportunities for Albanian chromites.
The last decade (1990-2000) was not so favourable for the chrome industry, with the exception of the year 1995 (see Metal Bulletin). The following table shows the Albanian chrome price record during 1990-2000:

TABLE VIII.

Prices (USD/ton)	*1992*	*1993*	*1994*	*1995*	*1996*	*1997*	*1998*	*1999*
1. Chrome Ore								
40-42 % Cr_2O_3	68.2	51.9	49.5	90.8	83.9	75.4	78.6	46.9
36-40 % Cr_2O_3	55.5	36.7	38.6	87.6	60.2	72.6	29.6	
30-34 % Cr_2O_3	28	27.3	28.9	90.8		51.9		
2. Concentrate								
48-50% Cr_2O_3	69.5		51.5	92.7	121.6	93.7		
45-47 %Cr_2O_3		43.06	50.3		113.5		86.03	67.3
3. Ferrochrome								
60-63%Cr	479.1	437.3	431.9	814.2	560.8	511.7	466.5	370.3

For the year 2000 the prices were roughly the following:
Ferrochrome was about 500 USD/ton; and
Chrome ore was 65 USD/ton (quality lumpy ore 40-42 % Cr_2O_3).
The year 2000 marks interesting figures in the Albanian chrome production and market.
Chrome ore produced by ALBKROM-state company — 60 000 tons.
Chrome ore produced by private mining entrepreneurs — 40 000 tons.
Ferrochrome produced by ALBKROM-state company — 15 000 tons.
Ferrochrome produced by DARFO (Italy) — 12 000 tons.
The private mining entrepreneurs sell chrome ore either to DARFO (Italy) or to foreign market (Sweden, Croatia, Belgium. Etc.).
There is *a tradition for more than 30 years* of trading chrome from Albania to Europe and the world, but now (starting from the second half of the year 2000) ALBKROM is not the only important trader. Within the year 2001 ALBKROM will be totally privatized.
The most of the trading is done "on the spot" or through short-term contracts.

INDUSTRIAL MINERALS AND SUSTAINABLE DEVELOPMENT IN GREECE

M. GROSSOU-VALTA and F. CHALKIOPOULOU

ABSTRACT: Besides the world-famous Greek marbles, Greece produces and processes a wide variety of industrial minerals and rocks. It is a leading producer of bentonite, second only to the USA, the world's third largest producer and exporter of perlite, the chief global exporter of pumice, a major magnesite trader and the sole global producer of huntite. This paper will give a comprehensive review of the evolution of the Greek non-metallic minerals industry and prospect of the future in terms of new fields of application.

1. Introduction

Mining activity in Greece is as ancient as the Greek civilization itself and besides metals, the exploitation of certain non-metallics is known to the Greeks since prehistoric times.

Figure 1. Ancient slate mine. V century B.C.

Industrial minerals of those times included obsidian for tools, sulphur for medicine and whitening of hair, plastic clays for ceramics, limestone and granite for construction and last but not least, marbles used for sculpture and construction purposes.
Processing was also developed by ancient Greeks. The best evidence being the Ancient Washery in Laurium Mines.

A.G. Tvalchrelidze and G. Morizot (eds.),
Mineral Resource Base of the Southern Caucasus and Systems for its Management in the XXI Century, 141–159.

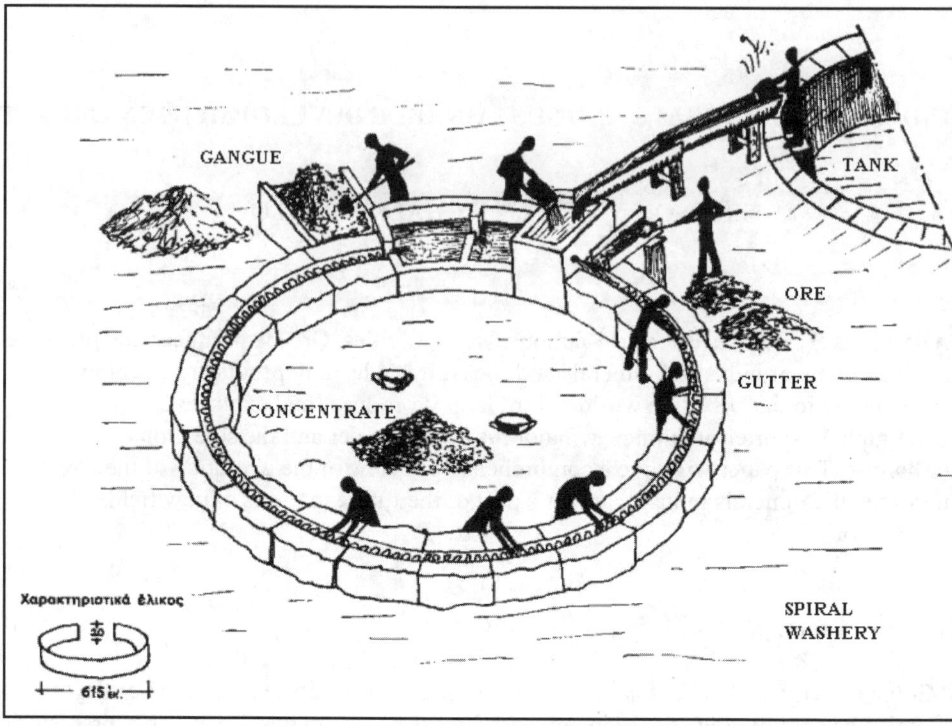

Figure 2. Ancient Greek processing installation

The mining and processing of industrial minerals was slow to develop. By the end of the XIX century there was limited exploitation of emery and pozzolana. The potential application of gypsum and magnesite deposits was also identified

The XX century marks the foundation of the first enterprises, activation of the extraction and processing of industrial minerals, mainly magnesite. The first rotary kiln for the production of dead-burned magnesia was installed in 1913.

Since then, there have been many developments in Greece and Europe in general due to the combined efforts of the private sector (mining companies and industrial mineral end-users), state research institutes and academic organizations.

Today production of industrial minerals in Europe significantly exceeds other ore production, the ratio being in the order of seven to one. The mineral sector in Greece is also quite dynamic and most of the mineral commodities produced are export-oriented. Thus in addition to marble, Greece produces, processes and exports a broad range of industrial minerals and rocks. It is the second largest bentonite producer in the world, and the world's third largest producer and exporter of perlite. Greece is the leading world exporter of pumice, a major magnesite exporter in EU, and the sole huntite producer in the world.

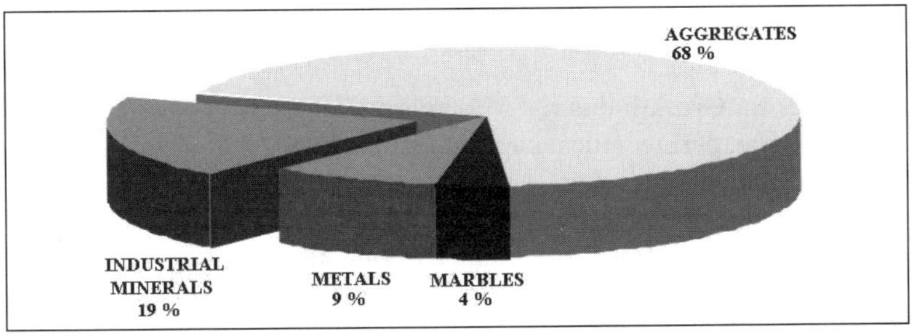

Figure 3. Production volume share (t)

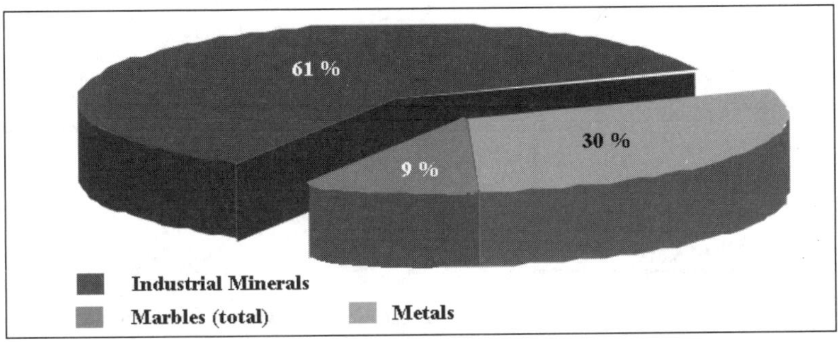

Figure 4. Export volume share

These minerals are used in traditional as well as in new areas such as environmental protection. Intense research, particularly in advanced processing and new technological developments, have created new markets for industrial minerals.

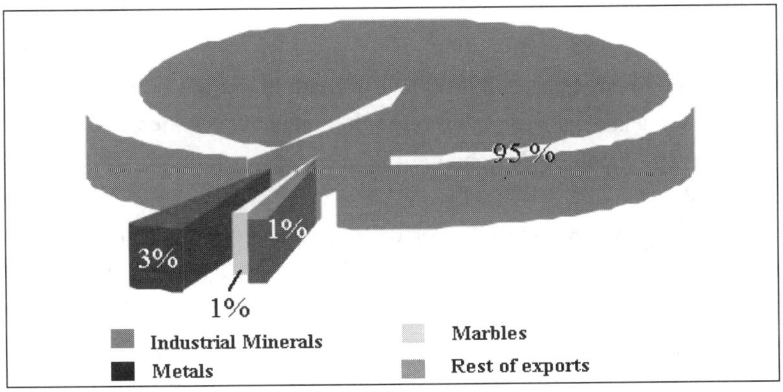

Figure 5. Export value share – Greek commodities

1.1. MAIN EXPLOITATION CENTRES

Important centres for Greek Industrial Minerals are:

- Milos (bentonite, perlite, silica, kaolinite, pozzolana).
- Yali island (pumice).
- N. Greece (magnesite, asbestos, feldspar, quartz, zeolites, huntite).
- W. Greece and Crete (gypsum, calcium carbonate).

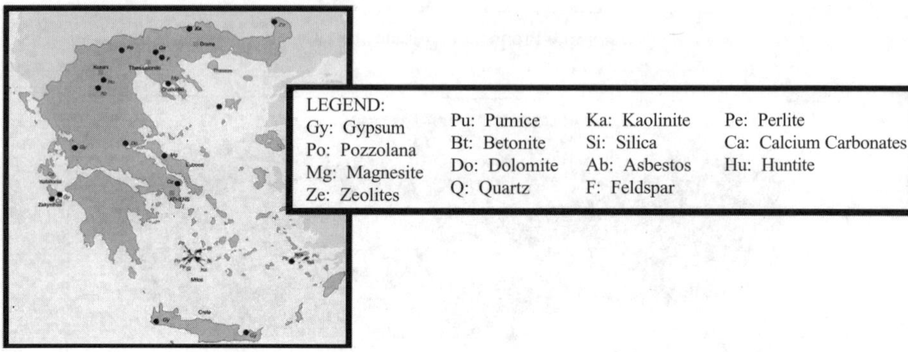

Figure 6. Industrial minerals in Greece. Location map with main exploitation centres

A very brief review of the main industrial minerals is presented in the following chapters.

2. The main industrial minerals

2.1. BENTONITE

Greece is the second largest world producer of bentonite after the USA, which is extracted from the island of Milos, although there has been some periodic extraction on the nearby island of Kimolos. Miloan bentonite contains predominantly Ca-montmorillonite **(>80%).**

There are two major bentonite producers which produce together more than 80% of the total output from Milos. Both companies use their own installations for the activation of bentonite with soda ash. A number of minor operators also extract bentonite in Milos. Virtually all bentonite produced is exported, mainly to the EU (77% of 1997 sales), North America and the CIS. Bentonite sales in the local market account for about 1% of the total sales.

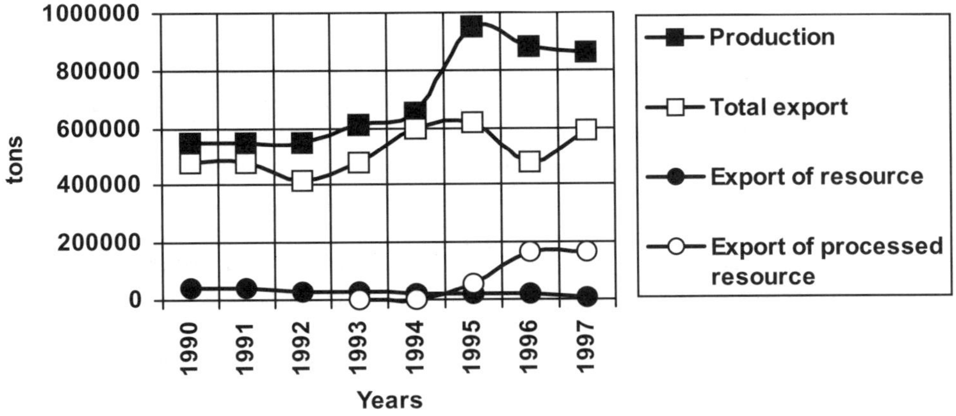

Figure 7.

As far as major applications are concerned, iron ore pelletizing accounted for 37% of the sales, foundries 33%, civil engineering applications 19% and drilling and cat litter markets 5% and 6% respectively.

Iron ore pellets: Bentonite serves as a binding agent in the production of spherical iron-ore pellets, which are suitable as feed material in blast furnaces, for pig iron production. Iron- ore pelletizing is directly connected to the steel industry and consequently bentonite sales are linked to the growth in the iron and steel sector.

Foundries: Bentonite is used in the traditional and most widely used method of sand mould casting. It serves as the bonding agent for the sand grains.

Civil engineering and environmental applications: There is a fast-growing use of bentonite in civil engineering applications, where it is used as a sealant in diaphragm walls, in foundations, as an additive in concrete, for the construction of tunnels and for the protection from leakage of underground and surface waters.

Drilling mud: Another conventional use of bentonite is as a mud constituent for oil and water well drilling. The role of bentonite is mainly to seal the bore walls, to remove cut material and to lubricate the cutting head.

Pet litter: Unlike the applications mentioned above, the use of bentonite in pet litter, concerns consumer rather than industrial product markets. Consumer preferences and promotion tactics by the dominant producers make the pet litter market considerably variable from country to country in Europe in terms of the type of a raw material used. However, due to its superior properties, bentonite has increased its share over the years in this growing market by substituting other absorbent materials.

Other applications: Bentonites are used as detergents, organoclays, geosynthetic clay liners, dessicants, in paper processing, although volumes in these applications are still low.

2.2. PERLITE

Perlite, a volcanic glass of rhyolitic composition, containing 2.5% of combined water, is an important Greek industrial mineral. Greece is the main exporter and the third largest perlite producer in the world. Most of the perlite produced is crushed, screened and dried, half of which is then exported. Only a small part of the screened material is consumed, in the expanded form, by the domestic market.
Greece is Europe's main supplier with competition from Turkey, Italy, Hungary and Armenia. The vast majority of the material sold is destined for expansion facilities in Europe, the east coast of the USA and the Middle East with small amounts also exported to the Far East.
Perlite is mostly extracted from Milos with total reserves of 1,000 m. tonnes.

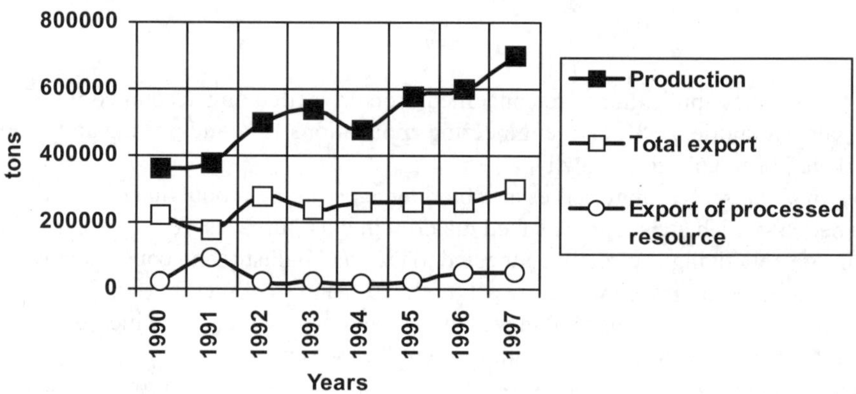

Figure 8.

Graded perlite products in excess of 500,000 tonnes were supplied in 1997 to international markets with a dominant position in Europe and a strong presence in North America. Virtually all material produced was exported mainly to EU (54% of the 1997 sales) and North America (41 %). Perlite sales in the local market account for 1.5% of total sales. Perlite from Milos and Kos is suitable for a variety of applications, mainly in the construction industry, but also in agriculture, filtration and in industrial insulation. Perlite can also be used as inert filler in plastics, glues, paints, explosives and varnishes. Construction and building materials account for the 66% of the total sales, 12% in agricultural applications and the remaining 12% for other applications such as filter aids, foundries and cryogenics.
The global perlite market is estimated to be around 2.1 m. tonnes, increasing at an

average rate of 4-5% per year, over the last five years. The market is geographically segmented with 39% in North America, mainly in the regions of New Mexico and Arizona, 31% in Europe, mainly in Greece (Milos), Italy, Turkey and Hungary and 25% in the Far East (mainly in China). Demand is expected to pick up further with increases in perlite consumption in applications other than construction, e.g. agriculture, production of filters, cryogenic applications, etc. In view of these forecasts, Greek perlite producers are channelling more investments into the mineral industry and are intensifying research.

2.3. MAGNESITE

The Greek magnesite industry shows upward trend since 1994. The main reason for this improvement is the restructuring of the sector with investments aiming at production cost cutting, improvement of the products and application of new marketing methods for the development of new markets.

Greek magnesite faced a hard period following the closure of Fimisco in 1991. Production fell from 900,000 tonnes in 1989 to 470,000 in 1992. Between 1991 and 1993, during the recession, prices slumped, picking up again in 1994. Ever since, there has been a marked improvement in demand as well as prices. An additional reason for this improvement is the antidumping measures imposed by the EU in 1993 in the form of minimum price for the Chinese dead-burned and caustic magnesia. Despite the need for better control and improvement of the efficiency of these measures, it is clear, that they are of vital importance and have substantially supported the EU magnesite sector. The export licence system introduced by the Chinese authorities, as a "by-product" of the antidumping measures, improved the market climate still further.

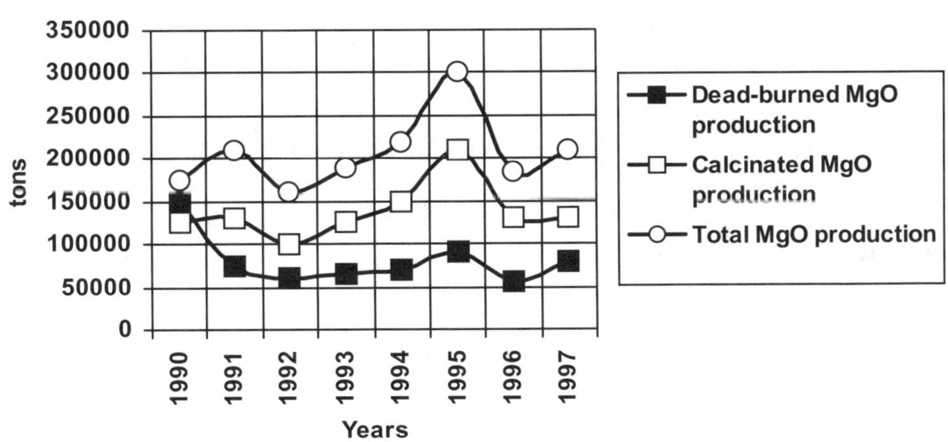

Figure 9.

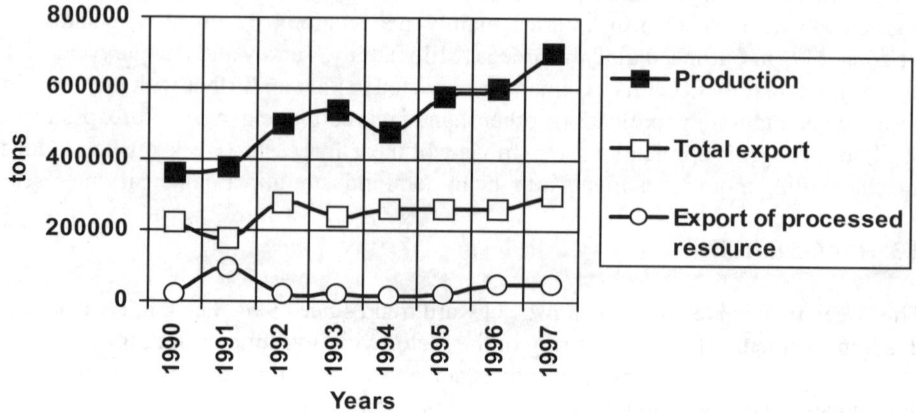

Figure 10.
Source: Grecian Magnesite SA

The end-products are dead-burned magnesia, caustic magnesia, raw processed magnesite and basic monolithic refractories. The total production capacity for fired products is close to 200,000 tpa. At least 96% of the output is exported. Within each of the above products, a wide range of grades is produced currently.
The sales value to EU countries is some 90% of its turnover while another 6.5% corresponds to the USA and the rest covers Australia, Bulgaria, Croatia, Hungary, Switzerland, and the Middle East countries.
The percentage of the sales volume destined to the domestic market ranged from almost 4% in 1995 to less than 3% in 1997. The small domestic market covers the following applications: basic monolithic refractories, abrasives and fertilizers.

2.3.1. *Main Applications*
$MgCO_3$: Ceramic, fertilizers, welding fluxes, slag conditioning, special applications.
Caustic MgO: Electrofused magnesia, electric-grade animal feed stuff and fertilizers, construction applications (magnesia floors, abrasives & panels), chemicals, environmental applications, Special applications.
Dead-burned: Refractory bricks and monolithics.
MgO: Welding fluxes, magnesium metal, special applications (electrical insulation).
Research in magnesite is progressing. In the last few years the biggest exploitation company has been engaged in research and development projects covering the following topics:

- Development of Mg compounds for flame retardant and other applications.
- Optimization of physical characteristics of caustic magnesia grades for

environmental applications such as neutralization of industrial acid effluents, heavy metal removal from effluent, flue gas desulphurization, municipal water supply treatment, etc.).

- Development of magnesia based heavy-duty clinkers correlation of performance of caustic magnesia grades employed for structural applications with their chemical and physical characteristics.
- Development of a new procedure for the exploitation of the 12 mm material.

Although the international markets for magnesite and magnesia based products are expected to be highly competitive over the next few years, there are good prospects for the further development of the Greek magnesite industry.

Key actions for the future development of the sector, for Greek magnesite producers and exporters are:

- To retain and further improve the quality of the penetration of end-products into new markets that have good prospects of development.
- Development of new applications.

2.4. PUMICE

Pumice is one of the most important industrial raw materials of Greece. It is a volcanic rock of extremely low density, with an off-white colour. The only exploitation site is located on Yali Island, in Aegean Sea, where there are extensive deposits of pumice with reserves exceeding 120 m tonnes.

There is only one pumice producer, operating a quarry on Yali Island since 1952. And it is the main producer and exporter of pumice in Europe and the leading exporter in the world. The raw material is only partly crushed and sized. Today the production consists of five different grades, varying only in their grain size analysis: 0-5 mm, 0-8mm , 5-8 mm, 8-16 mm and 20-70 mm.

In Greece pumice is mainly used in the construction sector for the manufacture of light-weight blocks, coating or insulation material. Additionally, being an abrasive material, it has many uses in the field of polishing metals, glass, plaster and leather. Finally, it is used in the textile industry for stone washing. Most of the production exported to the USA and EU is used for building purposes and only a very small part goes for stone washing. The latter is considered by the company as a declining market. The US market is important for the Greek producers, but somewhat uncertain due to the competition from domestic pumice substitutes and also due to freight rates. Small pumice markets in the Arab world and other countries are also developing. Lava pumice faces little competition in applications in the construction field and occupies the leading position in sales for the use of pumice blocks.

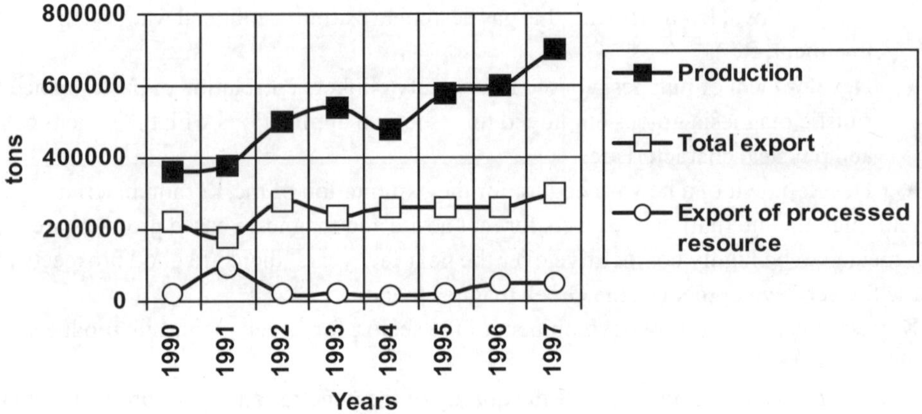

Figure 11.

Nevertheless, the Greek producer continues its research efforts in order to establish new applications, e.g. in agriculture for the neutralization of soils and in the manufacture of structural products with enhanced properties (light-weight, better sound and thermal insulation). The research for the latter application was recently successfully completed and the products developed are now being promoted to the market.

2.5. GYPSUM

Gypsum is produced mainly on Crete Island, Altsi deposit in the east, and Aetoloakarnania in the west (Katouna deposit). Periodically, minor production takes place in western Crete and in Zakynthos Island, in Ionian Sea.

Gypsum is of secondary origin, formed by the hydration of anhydrite. The extracted material is 85-90% pure with the impurity levels at the 10-15% mark (mainly SiO_2, $Fe2O_3$, Al_2O_3 and calcium and magnesium carbonates). Reserves are of the order of several tens of million tonnes.

The production of gypsum in Greece is largely dependent on the cement industry since it is the main consumer. The demand for gypsum in building applications is on the increase with the incorporation of gypsum into new materials with enhanced properties along with other industrial minerals such as vermicullite and perlite.

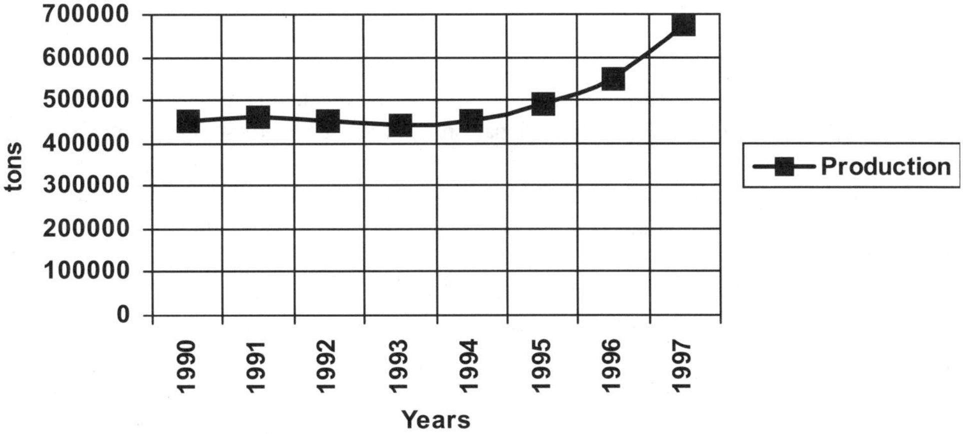

Figure 12.

Gypsum production in 1998 is estimated to be around 850,000 tonnes, 80% of which derives from the Altsi deposit, in Crete. Minor quantities were exported from Crete to other Mediterranean countries, for use in cement manufacture. Perspectives for 1999 will remain the same.

2.6. ASBESTOS

Despite the serious problems that asbestos market is facing internationally, the Greek asbestos business has managed to survive quite successfully, showing a steady increase in asbestos fibre production over the past few years.
The ore, located in Zidani, Kozani area of Northern Greece, is hosted in an antigorite serpentinite, which is formed under conditions of low temperature and relatively high-pressure metamorphism. These conditions favour the formation of chrysotile fibres. Proven reserves of 130 mt were initially calculated with 3% fibre content. Further potential reserves were identified which should extend the life of the mine to 25 years. Production is exported to Europe, Far East and Middle East, SE Asia and Brazil, mainly for the use in asbestos cement products.

2.7. BARYTES

There is only one Greek producer of barytes, with very limited production of the order of a few hundred tonnes per annum. The ore is extracted from different locations in the north-eastern part of the Milos Island. The reserves are estimated at 2.8 m tonnes with 58% $BaSO_4$ content.

2.8. FELDSPAR

The needs of the Greek ceramic industry in feldspathic products are in excess of 30,000 tpa and only a small part (mainly K-feldspars for porcelain) is covered by imports. The glass industry covers its needs (10,000 tpa) with imported Norwegian nepheline-syenite. There is only one Greek feldspar producer. Pegmatites composed of 70-75% feldspar, 15-25% quartz, mica and small concentrations of iron minerals, are the parent rock for the production of Na-feldspars.
Current research is focused in separating quartz from feldspar with the help of new cheap and environmentally friendly reagents.

2.9. QUARTZ

The combined quartz demand of the Greek ceramic and glass industries is around 80,000 tpa. The needs of the glass industry are covered exclusively by imports of glass sand.
Over the last few years the company has intensified its research efforts towards the development of advanced beneficiation techniques to produce ultra-pure quartz with SiO_2 content >99.99% for high-tech applications and the introduction of plasma technology for the manufacture of high-purity ferrosilicon.

3. Fillers (Calcium Carbonate & Huntite)

3.1. CALCIUM CARBONATE

Although calcium carbonate is, without doubt, one of the most widely used fillers in Europe, the involvement of companies in the production of white carbonate filler-grade products was relatively slow in Greece until 1993.
Zakynthos and Kefallonia, the two Ionian islands, are the main calcium carbonate production centres. Pure white friable microcrystalline limestone, dolomitic and calcitic marbles, as well as huntite (discussed separately), are the raw materials used for carbonate filler production in Greece.
There has been intense research aimed mainly at the processing of by-products obtained from calcitic marbles and aggregate quarry extraction with successful results in terms of the fillers industry. Specific research studies have been undertaken in certain areas, such as Drama, Kavalla, Thassos, important marble exploitation centres in Northern Greece. The brightness of the samples is generally very high and the inherent quality characteristics of the raw materials render them, in most cases, suitable for filler applications. On the other hand, very few companies (filler end-users) in Greece are aware of the benefits of incorporating mineral fillers in their production scheme. It is believed that this attitude will change with the help of proper market research.

3.2. HUNTITE

Huntite is an unusual carbonate mineral, $Mg_3Ca(CO_3)_4$, present in Upper Neogene lacustrine formations of the Kozani Basin, Northern Greece. To date the huntite of Kozani Basin is the only commercially exploited deposit in the world. In the Neraida-Lefkara region, southeast of Kozani, where the operation is located, the huntite deposit is composed of 95% huntite+hydromagnesite (in a ratio of 1:1). The remaining 5% consists of aragonite, dolomite, calcite and magnesite.
Two quarries operate in Neraida-Lefkara region producing huntite suitable for filler applications in polymers, where it acts as flame- retardant and reinforcement filler, paints, sealants, and high-quality papers (coating-grade). All the production of raw huntite and 80% of the processed material is exported. Until recently a domestic paper company utilized a small amount of the mineral as coating-grade filler in paper.
Total sales for 1998 reached 5,300 tonnes of raw huntite and 4,700 tonnes of processed and partly processed product. Perspectives for 1999 will remain the same. Currently the two producers are trying to develop new markets for their products.

3.3. DOLOMITE

Very pure dolomite, with MgO >20%, is quite abundant and widespread all over Greece. These dolomites are secondary products, derived from the dolomitization of Mesozoic limestones. Although the dolomite resources are in most cases suitable for a range of applications, including synthetic magnesia and refractory bricks, the ready availability of high-quality magnesite prevents much use in these applications.
A recent analysis of the Greek market showed that 37% of the industries, which could incorporate dolomite in the manufacturing process (paints, plastics and glass) ignore the potential of this industrial raw material.
Locally, the production manufactured by two minor operators in central Greece is mainly used in the domestic glass and ceramics industry, in agricultural applications (neutralization of soils, fertilizers, animal feed) and as a filler in paints. Impure dolomites and dolomitic limestones are used as a road aggregate and as railroad ballast. Recent laboratory research into certain dolomite deposits proved that dolomites with satisfactory abrasive behaviour and other relevant physical characteristics could be used as an alternative environmentally friendly sandblasting material.

3.4. KAOLIN

Kaolin is produced in Milos Island and Lefkogia in the Drama region of northern Greece. The two deposits differ in origin. Kaolin at Milos originated from the hydrothermal alteration of volcanoclastic rocks under acid conditions. These deposits are small, forming pockets or lenses and are controlled by structural criteria.
The main phase is kaolinite associated with quartz, cristobalite, alunite, barytes and Fe-

oxides and the following chemical composition: Al_2O_3=13-20% and Fe_2O_3= 0.3-0.6%. Its poor quality is mainly due to the opaline silica and alunite content. Reserves are estimated to be around 2.5 m tonnes.
Most of the production from Milos (80%) is consumed, in its crude form, by the domestic cement industry. The rest is used as filler in rubber and paper industry. A small amount (10%) is exported for the needs of white cement manufacture. The whole production from Drama region is domestically consumed by the ceramics industry, for the production of floor and wall tiles. The raw material is only crushed, ground and sized. Domestic industry requirements of high-quality kaolin are covered by imports. Since 1995 the kaolin production has drastically decreased due to depletion of reserves.

3.5. SILICA

Amorphous silica occurs on Milos Island, forming appreciable deposits only in two sites. The deposits contain more than 92% SiO_2, almost all of which is amorphous, with high reactivity, since it has a high specific surface area (25~30m²/g for granulometries below 65 ìm). Total estimated reserves are around 1 m tonnes. The mine is an open pit quarry and exploitation is being carried out using only bulldozers, without explosives. Processing is subcontracted in northern Europe and includes crushing, drying and micronizing in different granulometries.
At present there is no commercial production of amorphous silica and the material is currently being tested by the company for potential industrial applications. However, small volume sales have started in Northern Europe. Potential applications include water-glass (sodium silicate), calcium silicate, production of silica refractories, superpozollanas, speciality mortars, silicones, fillers and water treatment.

3.6. ZEOLITES

Natural zeolites may be defined as hydrated aluminosilicates of the alkali and alkaline earth metals, with an infinitely extended three-dimensional anion network and thus they have the atomic ratio O:(Al+Si)=2. Natural zeolites are remarkable for their continuous and partly reversible dehydration and for their baseexchange properties.
Several occurrences of zeolitic tuffs have been discovered and studied in Greece (islands of Milos and Polyaegos, Evros area) since 1981, when the first reference in the scientific literature to the exploitation potential of the zeolitic tuffs in Santorini Island was made. None of these occurrences had undergone exploitation until the beginning of 1996.
Zeolites are widely used in animal feed supplements, because of their high ammonium (NH_4+) binding ability, ammonia (NH_3) retention and moisture control. The zeolite product can be used as a feed additive for animals, for odour control in animal housing and as a free flowing agent to increase the speed of excretion during pelletization of feeds. The major advantage of the mineral when used in animal feed is the elimination of NH_4+ and NH_3. This results in the minimization of some of the environmental problems faced by farmers as well as an optimization of the hygiene conditions in farms.

The major export markets are in France, Germany, Benelux and the UK. Lower quantities have been sold for wastewater treatment and ammonia removal in fisheries. Demand is increasing since markets have become aware enough to accept this new product. Intensive research is being carried out in Greece nowadays especially in the sector of effective absorbents for the purification of water and pollution control.

3.7. MARBLES

Greece is a significant marble producer and well known from ancient times. Greek sculptures from white Penteli marble are exhibited in the museums of many countries and the Acropolis of Parthenon in Athens is one of the most celebrated monuments in the world.

Figure 13.

Nowadays Greek marbles are well known for their variety in quality and colour.

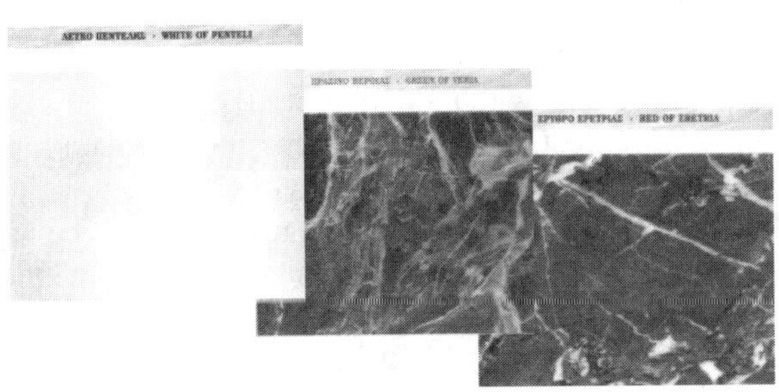

Figure 14. Marbles

4.000 both small and big enterprises with 60.000 employees work in the marble exploitation-workshops-placement and market.
GREECE HOLDS THE 5th position in world production. It is a very vital sector with a high potential in European and the International fields. The production is 2 million tonnes and the exports are about 110 million Euros.

A severe problem facing the European marble industry is the management of marble exploitation residues. A lot of research has been done on the environmentally friendly exploitation, management and deposition of this material. It has almost been proved that this is a 2nd generation material, which can effectively be used as filler and, in particular cases, a high-quality one. IGME is strongly involved in the field and runs several projects on the possibility of the use of marble residues as fillers for the paper industry, for paint, for plastics, for plasters, and also for the production of modular fuelling components and in general in the production of new low- cost materials for the construction industry.
It is believed that recycling of these 2nd generation materials will play a significant role in the industry of the XXI century in three directions:

- Environmental protection.
- Waste management.
- Low-cost raw materials.

4. Key Points in XXI Century Mining

- Progressive depletion of high- grade mineral resources.
- Environmental restrictions on the operation of mines.
- Participation of local communities in decision-making.
- Increasing needs for new high- quality products for demanding customers.
- Increasing international competition.

5. Sustainable Growth & Industrial Minerals

The field of industrial minerals and rocks is developing.
The trend is to combine growth with modern attitudes, i.e. integrated exploitation of ores, protection of environment, zero-waste processing, production of value-added materials, development of new uses, exploitation of residues.
Process technology plays a key role in industrial minerals, particularly in the production of high-value products. Process technology can also provide the means to develop specific properties so as to introduce improved or new products into the marketplace and thus increase market share as well as profitability. Besides the impact of increasing market value of products, new technology can utilize the resources of previously unusable minerals thus expanding the reserve base of a company. Other process technologies not affecting the product can lead to the reductions in production costs.
In view of new technological developments and international competition, the industrial mineral wealth of Greece needs to be re-evaluated and new research targets to be set, taking into account the requirements of modern society, industry, energy conservation and environmental protection.
In response to technological development, international competition and environmental

legislation, the Greek industrial minerals sector has set up various research and development projects aiming at:

- Development of new products and markets. Development of new fields of application.
- Production of low-cost, value-added products from unexploited resources, rejects, or by-products.
- Use of low-cost industrial minerals in environmental protection.
- Use of new technologies and the development of new methodologies to upgrade mineral resources.
- Minimization of production cost.
- Minimization of environmental impact.

Process technology plays a key role in the production of high-value products and in the introduction of improved or new products into the marketplace.
New technologies can utilize previously unusable mineral resources and may result in reduced production costs.
Benefits obtainable with different processing technologies are presented in the following table:

	High- Value Products Production	*New / Improved Reserves*	*Extension of*	*Cost Reduction*
Pulverization/ Classification Systems	X	X		X
Dispersion Systems	X	X		X
Wet Classification	X	X	X	X
Froth Flotation	X	X	X	X
High-Intensity Magnetic Separation	X	X	X	
Media Grinding	X	X	X	
Chemical Surface Treatment	X	X		
Chemical Surface Treatment	X	X	X	X
Selective Flocculation Processes	X	X		
Ozonation – Bleaching			X	X
High Solids Slurry Evaporation		X		
Chemical Structuring Processes	X	X		

6. New Products - Example

Modern technology including size reduction and mixing with resins applied on cheap raw materials may result in new construction products.

Figure 15. The raw material

Figure 16. Synthetic marble

Ultra-fine grinding (jet milling) may transform quarry rejects into high-value fillers for modern applications.

Figure 17. Quarrying rejects

Development of industrial minerals is based on 2 critical points: Market and environment.

1. Fulfilment of modern market demands refers to:
- Production of standard high- quality, tailored raw mineral materials with enhanced properties that increase the price of the final products.
- Production of ready mixes.
- Development of new / improved products in order to substitute given materials or enter new marketplace.
2. Environmental protection has to do with:
- Preservation of mineral resources – Extension of reserves.
- Water-energy preservation.
- Recycling.
- Land reclamation.
- Improvement of production procedures.

7. Proposals

1. Support R&D projects that include scale-up in order to:
- Incorporate innovative, environmentally friendly, low-cost technologies in industrial enterprises dealing with the exploitation of industrial minerals.
- Improve the competitiveness and the viability of the relevant industry.
2. State support of the Industrial Minerals' Industry with specific financial measures:
- Application of new technologies.
- Extension of their activities by e-commerce as well.
- Modernization of installations in accordance with market demands.

Having in mind all the presentation so far, we come to the conclusion that promotion of the co-operation in the sector of industrial minerals between the EU and Caucasian countries, which have close interests and relative civilizations, should be started in this workshop, for it is the best opportunity therefor.

8. References

1. Grossou-Valta, M., Chalkiopoulou, F. (1998) The Greek Industrial Minerals and their applications in local and foreign markets, *Greek Ceramic Company Conference*, Athens.
2. Hatjilazaridou, K., Grossou-Valta, M., Chalkiopoulou, F. (1998) *Greek industrial minerals*, current status and trends, *Industrial Minerals (June 1998)*.
3. Grossou-Valta, M., Chalkiopoulou, F. (1998) The value added to Industrial Minerals after Processing with New Technologies, *Workshop with the title: "New techniques in the aggregates' quarrying industry of Cyprus"*, Nicosia, Cyprus *(21 November, 1998)*.
4. Grossou-Valta, M., Chalkiopoulou, F., Kaklamanis N. (1999) Development of novel processing for the production of low-cost by-product fillers as a replacement for high-cost primary fillers, (REFILL) *Mid-Term Report*.
5. Grossou-Valta, M., Chalkiopoulou, F. (2000) Technology, a tool for the increase of the mineral ore reserves and the elimination of the environmental impacts, *3rd Congress of Mineral Wealth,* Greece, Athens *(November 2000)*.
6. Chalkiopoulou, F., Grossou-Valta, M. (2001) The prospects for the exploitation of the European White Marbles as Raw Materials for Fillers' Production, under edition, volume is dedicated to A. Frangiskos, the NTUA Professor *(April 2001)*.

MINERAL RESOURCES IN PORTUGAL - AN OVERVIEW

M. L. P. DE AMORIM DINIS and A. BOTELHO DE SOUSA
Laboratory of the Geological and Mining Institute of Portugal,
R. da Amieira, 4465 S. Mamede de Infesta, Portugal
Phone: 351 29511915 Fax: 351 2 9514040
E-mails: maria.dinis@igm.pt and ana.botelho@igm.pt

ABSTRACT: An overview of the Portuguese mineral resources is done in this paper. These resources are classified in subsectors mainly as metallic and non-metallic ores, industrial and ornamental stones and natural mineral and spring waters.
The subsector of metallic ores has been decreasing in importance, similar to what is happening world-wide. However, due to their relevance, three deposits should be mentioned, namely Neves-Corvo and Aljustrel, in the Iberian Pyrite Belt, and Panasqueira, in the Centre of Portugal.
The non-metallic ores include industrial minerals for the ceramic industry, such as clays from sedimentary deposits, feldspar from pegmatite and quartz from sand formations. Non-traditional new resources, such as arkoses and spodumene have been studied in order to meet the requirements of the glass and ceramics industries.
The Industrial Stones are mainly sands and building materials. The most important ornamental stones in Portugal are granites, marbles, and limestones. The production in these two subsectors continues to have a tendency to rise. An increase from 60% to 64% was verified in the period of 1996 to 1999.
The natural mineral and spring waters constitute a great potential in this field evidenced by the large number and the wide variety of occurrences. In the last few years this subsector shows the tendency to increase.

1. Introduction

The mineral raw-materials are non-renewable resources and those of better quality are being quickly depleted, so it becomes more and more necessary to preserve them for nobler ends, and to treat those of inferior value, in order to improve their quality. Moreover, the globalization of the market increased the number of demanding consumers, who will not tolerate products without quality, as well as of foreign suppliers with capacity to meet the required specifications. All these prerequisites force the producers to look at the possibility of having to valorize their raw materials.

A.G. Tvalchrelidze and G. Morizot (eds.),
Mineral Resource Base of the Southern Caucasus and Systems for its Management in the XXI Century, 161–166.

In the case of metallic ores, most of the mines, which were still in operation until the beginning of the eighties, were closed down. Only the high grades and tonnages of Neves-Corvo (massive sulphide, bearing mainly Cu and Sn) and Panasqueira (hydrothermal deposit, mainly containing W, Sn and Cu) mines, allow their current labouring, although being also dependent on external factors.

From the mid-eighties, a decrease in the production of the mining subsector was registered, as a consequence of the fast regression that was verified in metallic, precious and energy ores. This was caused by a decrease in the international quotation of precious and base metals and the reduction in the production of energy minerals: coal and uranium. In the case of coal, the extreme consequence was the closing down of Pejão mine, situated near Porto, in the North of Portugal. In the case of uranium it was the decrease in its importance, as an energy mineral, that contributed to the reduction of this subsector.

As for the ornamental stones, industrial minerals and non-metallic ores, due to their more common occurrence at the surface or at low depths, they could, in most cases, be used in industry as a run-of-mine material or after rather insignificant beneficiation.

2. Metallic ores

Due to their relevance, three deposits should be mentioned in the subsector of metallic ores: Neves-Corvo, Aljustrel and Panasqueira.

Neves-Corvo (>270 Mt.) and Aljustrel (>220 Mt.) are the most important deposits occurring in the Portuguese sector of the Iberian Pyrite Belt (I.P.B.), a massive sulphide situated in the South region of the Iberian Peninsula extending both in Spain and Portugal.

Now, Neves-Corvo is the only Portuguese mine in operation at the I.P.B. It was discovered in 1977 and opened in December 1988. Currently the main ores extracted are copper from chalcopyrite and, since 1990, tin from cassiterite. The concentrates of these two ores are destined essentially to exportation. Neves-Corvo mine has already become the biggest producer of copper and tin in Europe. In 1994 the mine produced 612.4×10^3 t of copper and 7.77×10^3 t of tin concentrates. Total reserves of 270.1 Mt. grading 7.94% Cu, 1.41% Zn and 2.42 Sn% were estimated for the five ore bodies discovered so far (Neves, Corvo, Graça, Lombador e Zambujal). The fall of the copper international quotations verified in the last two years, caused a production decrease in this mine and consequently reduced the relative importance of the mining subsector of this region.

The Aljustrel mine, also situated in Iberian Pyrite Belt, had a great importance as a Zn, Pb and Cu producer. The Aljustrel deposit is also formed by five ore bodies (Algares, Moinho, Feitais, Estação and Gavião). For technical and economical reasons its exploitation was suspended in 1991, being now under economic feasibility appraisal.

Besides Neves-Corvo and Aljustrel Mines, prospecting and research works carried out in this region confirm the existence of potential resources with great value in zinc, copper, gold and silver, for its being necessary to really consider the joint exploitation of these resources in the future.

Panasqueira is a wolframite mine, situated in the Centre of Portugal. It also produces cassiterite and chalcopyrite as by-products. During 1994 this mine suffered a great crisis due to the fall of WO_3 international prices, which resulted in its closing down. The mine reopened in 1995. After the mine has restarted operation, about 40 000 tons of run-of-mine containing c.a. 0.38% WO_3 are treated per month. Despite the continuously low WO_3 quotations, this ore deposit is of world-wide importance, due to its grades and tonnage; it is the only wolframite mine still opened in the European Union.

3. Non-metallic ores

The traditional sources of non-metallic minerals in Portugal are mainly clays, including kaolin from sedimentary deposits, feldspar from pegmatite and quartz from sand formations. These resources are lacking both in quality and quantity, so it became necessary to look at other types of deposits, which may be rendered suitable for application in this branch of industry after their beneficiation. Their valorization demands higher levels of technological processing. Two specific cases should be pointed out:

- The 35 Mt. K-feldspar rich Arkosic ore deposit (a natural mixture of feldspar, quartz and kaolin), situated in the Centre of Portugal. In order to achieve the characteristics required by glass and ceramics industries, studies on the Arkosic ore deposit were conducted that included estimation of reserves and of a possible treatment diagram from which four products could be obtained: a rich feldspar concentrate (flotation), a feldspar/quartz mixture (magnetic separation), kaolin and sand for glass.
- The spodumene ore (a lithium-bearing ore containing also K-feldspar, quartz and muscovite) occurs in aplite-pegmatitic lodes in granites located in Northern Portugal. A study was conducted, envisaging its beneficiation, in order to obtain lithium concentrates for addition to glass and ceramic bodies. High-grade concentrates assaying more than 7% Li_2O (flotation) and glass-grade concentrates at c.a. 5% Li_2O (heavy media separation) could be obtained. A study of treatment diagrams was also carried out.

The industrial importance of this study is based on the valorization of the above mentioned 35 Mt. K-feldspar rich arkosic ore deposit.

4. Industrial and ornamental stones

The ornamental and industrial stones have been evidencing a significant growth mainly due to the increasing competitiveness within the enterprises as a consequence of their valorization and higher quality. Most of the ornamental stones are produced in Alentejo, a region in the South of Portugal, which is the most important area for ornamental

marble and granite extraction in the country. In the North there is also a significant production of ornamental granite, appreciated for its high quality and beauty. The granites represent the most important rock outcropping in the country: there are more than 50 commercial designations, ranging from whitish to pink, greyish and dark grey. The extraction of black schist and serpentine to the ornamental ends in the North of Portugal should also be pointed out, which is the only region in the country where these substances occur.
The limestone is exploited mainly in the Central region, where 700.000 tons were extracted in 1998. This is a common resource in all country having more than 30 different commercial designations.
Concerning the industrial stones, new treatment technologies, implying a higher quality control production of sand and building materials, have been implemented. These technologies have the purpose to minimize the production of residues and to obtain different final products from the same ore deposit. This subsector has significantly increased in the last years, reflecting the increase in consumption of these raw materials by the construction sector. The greatest consumption of these materials takes place in the North and Centre (Lisbon) of Portugal.

5. Natural mineral and spring waters

The mineral and spring waters, whose occurrence in Portugal is favoured by the geological conditions, constitute a great potential as a mineral resource, evidenced by the large number of occurrences and the great hydrochemical diversity, which provide a variety of chemical compositions and temperatures at the source. These occurrences can be grouped under following categories: still waters with very low mineralization, widely used for bottling; natural sparkling waters with sodium bicarbonate, also used for bottling; and sulphurous waters, which are greatly used in spas on account of specific therapeutic properties.
The Northern and Central regions comprise about 74% of the hydrothermal occurrences and known spring and mineral waters. They support two important industries today, thermalism (spa) and water bottling. These subsectors have evidenced a significant growth in the last years. The great number of spas, directly related to this sector, has been relatively stable. It should be pointed out that the thermal activity is responsible for an important increase in national tourism.

6. Production in the mining industry from 1996 to 1999

The four following figures show the production by subectors in the Mining Industry, from 1996 to 1999.

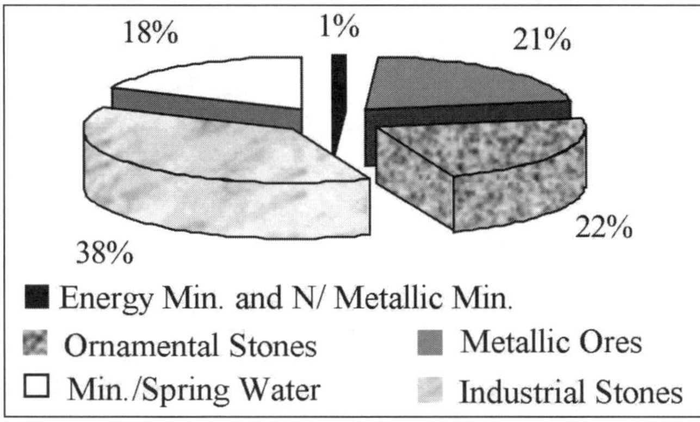

Figure 1. Structure of Value of Production by Subsectors, 1996

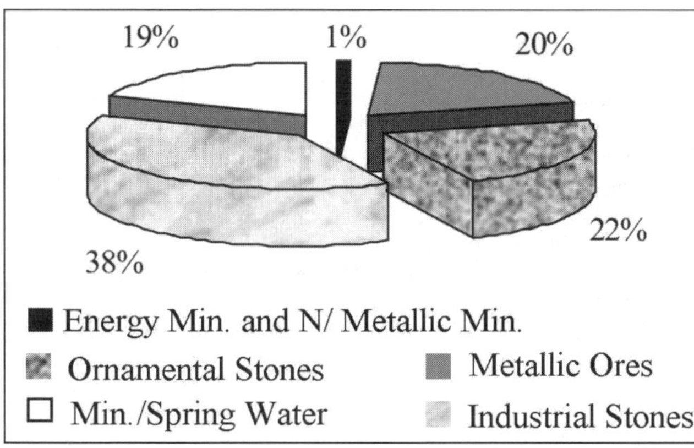

Figure 2. Structure of Value of Production by Subsectors, 1997

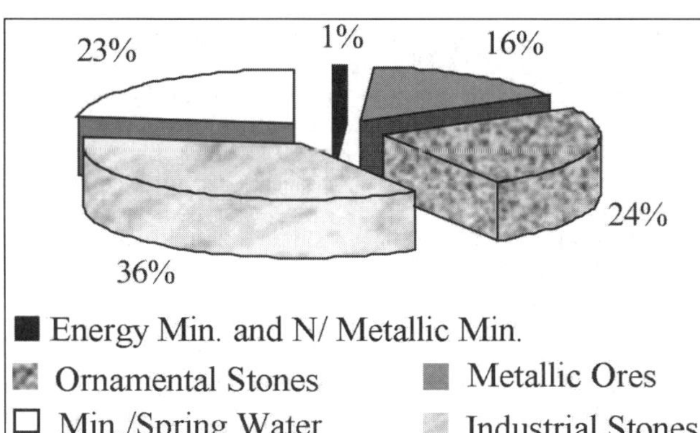

Figure 3. Structure of Value of Production by Subsectors, 1998

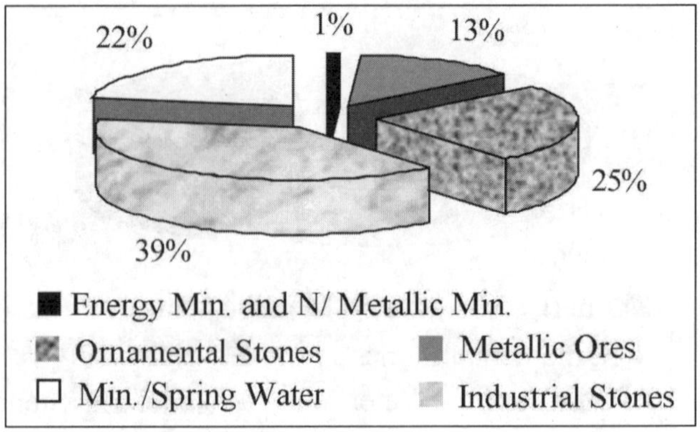

Figure 4. Structure of Value of Production by Subsectors, 1999

7. References

1. Amarante, M.M., Botelho de Sousa, A. and Machado Leite, M. (1997) Beneficiation of a Feldspar Ore for Application in the Ceramic Industry, SAAIM, pp. 193-196, (Julho/Agosto 1997).
2. Amarante, M.M., Botelho de Sousa, A. and Machado Leite, M. (1999) Technical Note: Processing a Spodumene Ore to Obtain Lithium Concentrates for addition to glass and Ceramic Bodies, Minerals Engineering, **12**, No 4, pp. 433-436.
3. Botelho de Sousa, A., Amarante, M.M. and Machado Leite, M. (2000) Beneficiation Studies on a Spodumene Ore from Portugal, *Proceedings of the XXI International Mineral Processing Congress, Rome, Italy, (July 2000)*, pp. C11-40 — C11-46.
4. Brochuras: Instituto Geológico e Mineiro (1999, 2000) Indústria Extractiva em Portugal, *Informação Estatística N º 5 (1999) e N º 6 (2000).*
5. Brochures: Geological and Mining Institute (1997, 1998) Mining Industry in Portugal, *Statistical Information No 3 (1997) and No 4 (1998).*
6. Carvalho, J., Manuppella, G. e Casal Moura, A. (2000) Calcários Ornamentais Portugueses, *Boletim de Minas*, **37**, Nº 4.
7. Cláudio dos Reis, A. (1991) As Minas da Panasqueira – Monografia, *Boletim de Minas, Lisboa,* **8**, N.º 1, pp. 3-44, *(Jan/Mar 1991).*
8. Correia de Sá A., Naique, R.A. and Nobre, E, (1999) Minas da Panasqueira - 100 Anos de História Mineira, *Boletim de Minas 36*, pp. 3-22, *(Jan/Mar 1999).*
9. Gaspar, O.C., (1996) Microscopia e Petrologia de Minérios Aplicados à Génese, Exploração e Mineralurgia dos Sulfuretos Maciços dos Jazigos de Aljustrel e Neves Corvo, *Estudos Notas e Trabalhos - Instituto Geológico e Mineiro,* Tomo **38**, pp. 3-189, *(Porto 1996).*
10. Grupo mineralogista de Madrid (1997) Bocamina Panasqueira, *Revista de Minerales e Yacimientos – Especial Internacional Portugal.*
11. Instituto Geológico e Mineiro (2000) CERÂMICA – Indústria Sustentável em Matérias-Primas Nacionais? *Encontro promovido pelo IGM e pela APICER no Laboratório do I.G.M., (Junho de 2000).*
12. Instituto Geológico e Mineiro (2000) Granitos e Rochas Similares de Portugal, *(Edição 2000).*
13. Instituto Geológico e Mineiro (2000) Indústria Extractiva.
14. Machado Leite, M.R. (2000) As Pedras e os Minerais nos Caminhos da História, *Comunicação, XIII Congresso da Ordem dos Engenheiros, Porto, (Setembro de 2000).*
15. SOMINCOR (1992) Neves-Corvo Mine.

METALLOGENIC & ENVIRONMENTAL INFORMATION SYSTEMS: A MODERN TOOL FOR THE SUSTAINABLE DEVELOPMENT OF MINERAL RESOURCES

The "GIS Central Europe" example

D. CASSARD and Y. ITARD
BRGM, Mineral Resources Division
BP 6009, 3 av. Cl. Guillemin, 45060 ORLEANS Cedex 2 – France.
Correspondence: d.cassard@brgm.fr

KEYWORDS: Central and South-eastern Europe / GIS / Metallogeny / Ore deposits, Metal potential and fertility maps / Geodynamics / Environment / Risk maps.

1. Introduction

The extractive industry, for many Central and South-eastern European countries, has long been an essential component for industrial development, and geology (in its broadest sense) a discipline of the utmost importance. Today, the entire industrial sector in Central and South-eastern Europe must adapt itself to a much more competitive economy. Current problems in this area can be attributed to a combination of several factors such as entry into the global market, the emergence of "previously non-existing" environmental constraints, lack of capital, an inherited non-competitive industry, and slow adaptation of the administrations to their new role of "controller".
Reorganization of the mining sector in these countries would nevertheless benefit from favourable factors such as a large amount of existing geological work, the high quality of the accumulated mineral-exploration data, and the availability of highly qualified and relatively low cost labour. Mining activities in these regions, whether the mines be large or small, can still be a major development factor generating sustainable employment and income, *but only if they take into account the parameters adopted by the mining industry internationally, along with what is now a necessary environmental approach.*

2. How can a Metallogenic & Environmental Information System help in establishing sustainable development policies?

In the present case it is a question of developing a Geographic Information System (GIS) based on geological and industrial data related to the mines and environments of Central and

A.G. Tvalchrelidze and G. Morizot (eds.),
Mineral Resource Base of the Southern Caucasus and Systems for its Management in the XXI Century, 167–180.

South-eastern Europe. The purpose of this GIS (termed "GIS Central Europe", for brevity) will be to produce *decision-aid documents that integrate all the necessary factors related to industrial activities, and in particular, the extractive industry, for medium- to long-term regional development planning and ecosystem protection.* GIS Central Europe will be a tool for national, regional and European authorities, as well as for industrialists. It will also be useful as a *training* tool for those involved in this sector, and as a *communication vector* for data dissemination and access. The proposed documents will be:

(i) Maps of economic development potential for the mining and other sectors;
(ii) Databases of environmental risk factors related to different activity domains in the industrial sector at regional scale;
(iii) Decision-aid documents combining mining, economic, social and environmental aspects.

The anticipated downstream benefit from this project is a solid-based adaptation, modernization and reorganization of the Central and South-eastern European mining industry that will, in turn, lead to the development of other related activities. This should, in principle, be accompanied first by a general rise in living standards for those involved in the mining industry, whether directly or indirectly, and subsequently by an increase both in trade and in technical and scientific cooperation between the countries of Central/South-eastern and Western Europe.

3. The components of GIS Central Europe

A GIS is a highly sophisticated tool that has now become the accepted method for processing information from multiple heterogeneous datasets. Once these datasets have been entered into the system as geo-referenced thematic layers (Cassard, 1999 a, b [2,3]), the data can be combined in any way the end-user sees fit for resolving specific problems.

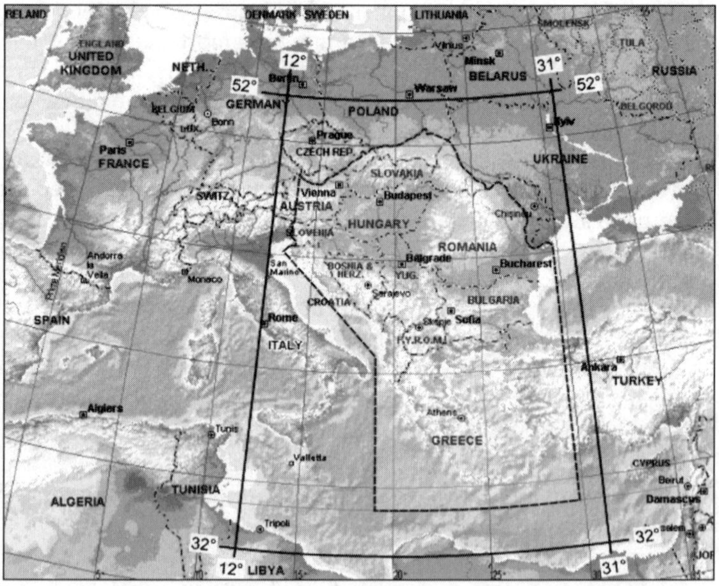

Figure 1. Area covered by GIS Central Europe between lat. 32° to 52° N and long. 12° to 31° E.

GIS Central Europe (Figure 1), drawing heavily on the authors' experience in creating GIS Andes (Cassard 1999b,c [3,4]) and GIS Gondwana (Nicol, 2000 [12]), will comprise three modules – "Geography", "Geology & Mines" and "Environment". The thematic layers to be contained in the first two modules are given in Table I:

TABLE I. Thematic layers of the GIS Central Europe "Geography" and "Geology & Mining" modules.

Layer	*Description*
GEOGRAPHIC	A DCW® geographic base
DEM	Two digital elevation models: one oceanic (2-minute arc), and one continental (30-second arc) elevation dataset with a structural analysis of the detailed topography
IMAGERY	SPOT 4 VEGETATION® satellite images at 1x1 km (Figure 2)
GEOLOGIC MAP COVERAGE	Present state of geological coverage: location of and information on the existing maps
GEOLOGIC SYNTHESIS	Synthetic geological map of Central and South-eastern Europe at 1:1,500,000 scale
SIMPLIFIED GEOLOGICAL MAP	Simplified geological map of Central and South-eastern Europe at 1:1,500,000 scale with characterization of morpho-structural domains and main tectonic elements (see Tomek et al., 1996 [15])
VOLCANIC	Data on Holocene volcanism
GEOTHERMAL RESOURCES	Based on inventories at present being compiled for the EU
GEOCHEMISTRY	Composition and age of magmatic and volcanic rocks – isotope data
ORE DEPOSITS	Linked to a database under Access®, using a new metallogenic lexicon (see http://giseurope.brgm.fr/) and including mineralogical (ore, gangue and hydrothermal alteration), fluid-inclusion and isotopic data of the main ore deposits (see, for example, http:// www.brgm.fr/sigand/mineralo.htm) (Figure 3)
GEOCHRONOLOGY	Synthesis of existing data, methods used, critical analysis, reliability, references, etc.
MINING DISTRICTS AND PROVINCES	Delimitation, magmatic and structural controls, potential and information on mining companies (see Jankovic, 1997 [9])
HEAT FLOW	An up-to-date synthesis (see also Cermak, 1993 [5])
GRAVIMETRIC	Bouguer anomaly calculation; isostatic correction and corresponding residual anomalies; vertical gradient calculation and structural analysis
SEISMIC	Distribution of earthquakes in order to better understand and constrain the crustal structures
MOHO	Depth contour map (e.g. Nemcok et al., 1997 [11])
3D TOMOGRAPHY	3D model of the lithospheric structure (see de Boorder et al., 1998 [6])

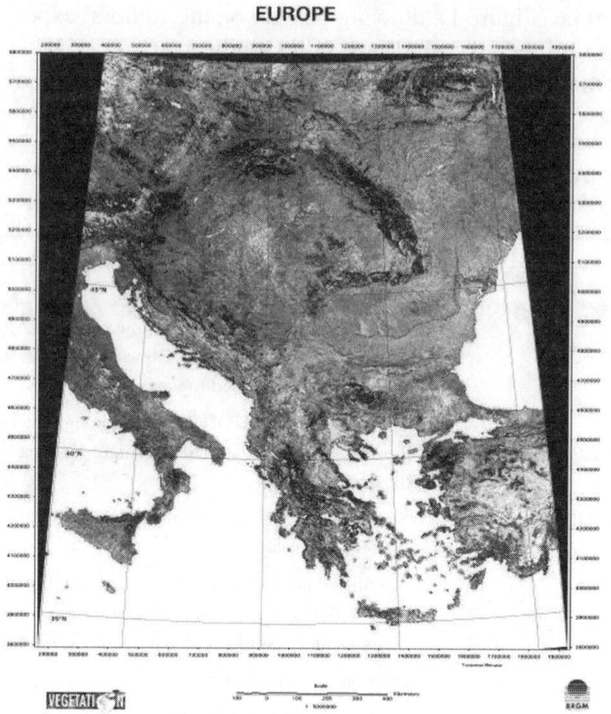

Figure 2: Ten-day global synthesis satellite image of Central and South-eastern Europe from the CNES – VEGETATION Programme.
Resolution: 1 km – TM projection: Spheroid Clark 1866 – Processing: BRGM

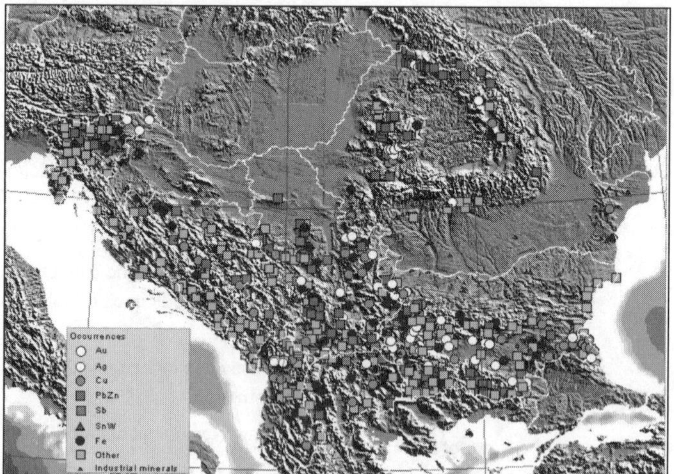

Figure 3: GIS Central Europe "Ore deposits" layer (still under construction).

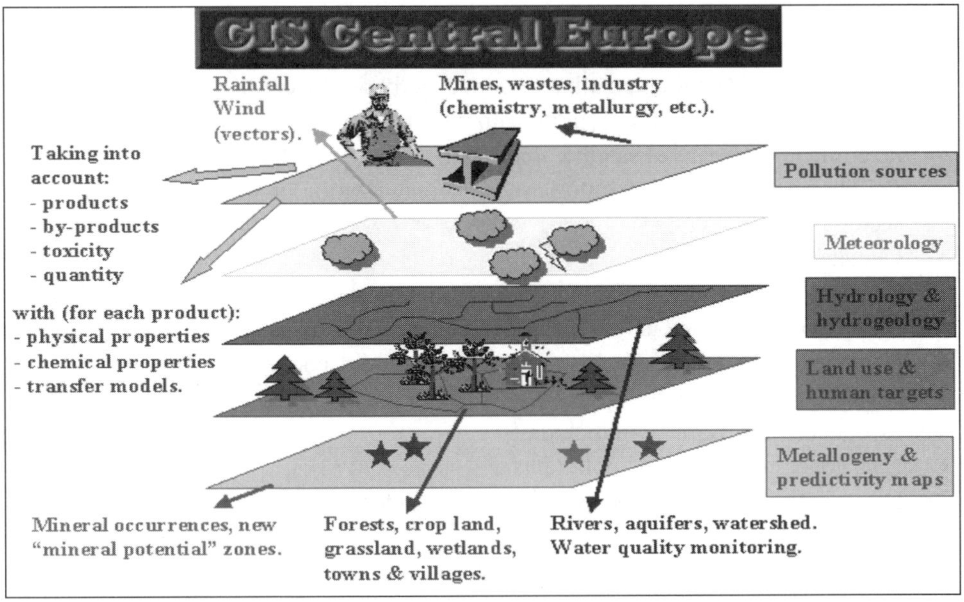

Figure 4: Layers of the GIS Central Europe Environment module.

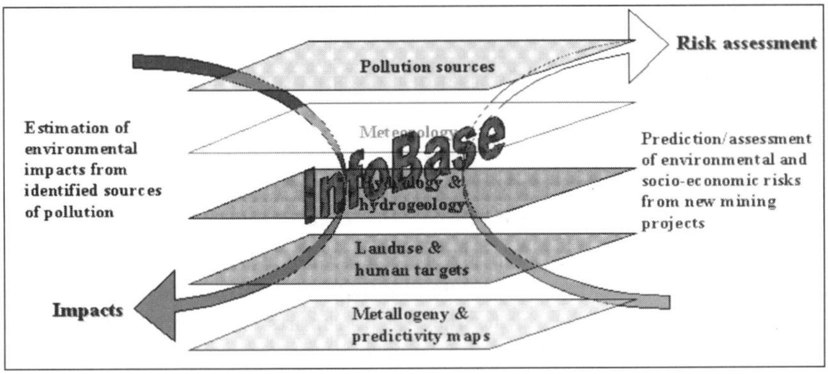

Figure 5: Two areas of application for the GIS Central Europe Environment module.

The Environment module of GIS Central Europe will include a descriptive section of the different environments and sectors that will be linked to an "InfoBase" of data on potential pollutions, activity sectors using potentially polluting products, and relevant national and international regulations and standards.

The descriptive section of the Environment module, at present made up of five layers — Pollution sources, Meteorology, Hydrology & hydrogeology, Land use & human targets, and Metallogeny & predictivity maps (Figures 4 and 5) — will provide information on:

- Detailed Administrative boundaries showing environmentally protected areas such as National Parks;
- Distribution and characterization of human activities (socio-economic aspects);
- Land use (agriculture, forest, urbanism, industry, mine waste, *etc.*), mainly determined by means of satellite imagery;
- Pollution sources (with inventories and characterization) in the different activity sectors;
- Climate;
- Water resources (both surface water and groundwater);
- Risk vulnerability and natural hazards (erosion, earthquakes, flooding);
- Geothermal resources (*cf. above* – role of alternative energy resources);
- Quality monitoring networks for water, air, and soil.

The associated "InfoBase" will include:

- Data concerning potential pollutants and associated risks;
- Data on the activity sectors: products used, products manufactured, by-products generated;
- Data on European regulations (laws, standards).

4. Development of a GIS

The construction of a successful Geographic Information System relies essentially on the quality of the data that it contains, i.e. on the intrinsic quality of the data and the way in which it is represented and archived in the base. For example, the quality of the data stored in an Access® type relational base, as is the case for GIS Central Europe, is dependent to a great extent on the quality of the lexicons used for describing the data. The full development of a Metallogenic GIS generally takes a number of years (depending on the allocated resources) and includes several stages:

- Stage 1: An exhaustive search for data in old reports. The work carried out during this phase is used for establishing the GIS base layers, i.e. the information layers on existing data that we call metadata (e.g. Mapping Coverage - Figure 6);
- Stage 2: Data compilation and quality evaluation; elimination of doubles and selection of data to be retained preferentially; homogenization, organization, and constitution of the so-called "factual" layers (e.g. Heat flow, Gravimetric, etc. - Figures 7a and b);
- Stage 3: Use of the selected data to generate summary documents. This requires a significant amount of know-how and experience and provides "synthesis" or "added value" layers of the "Geological synthesis" type (Figures 8a and b);

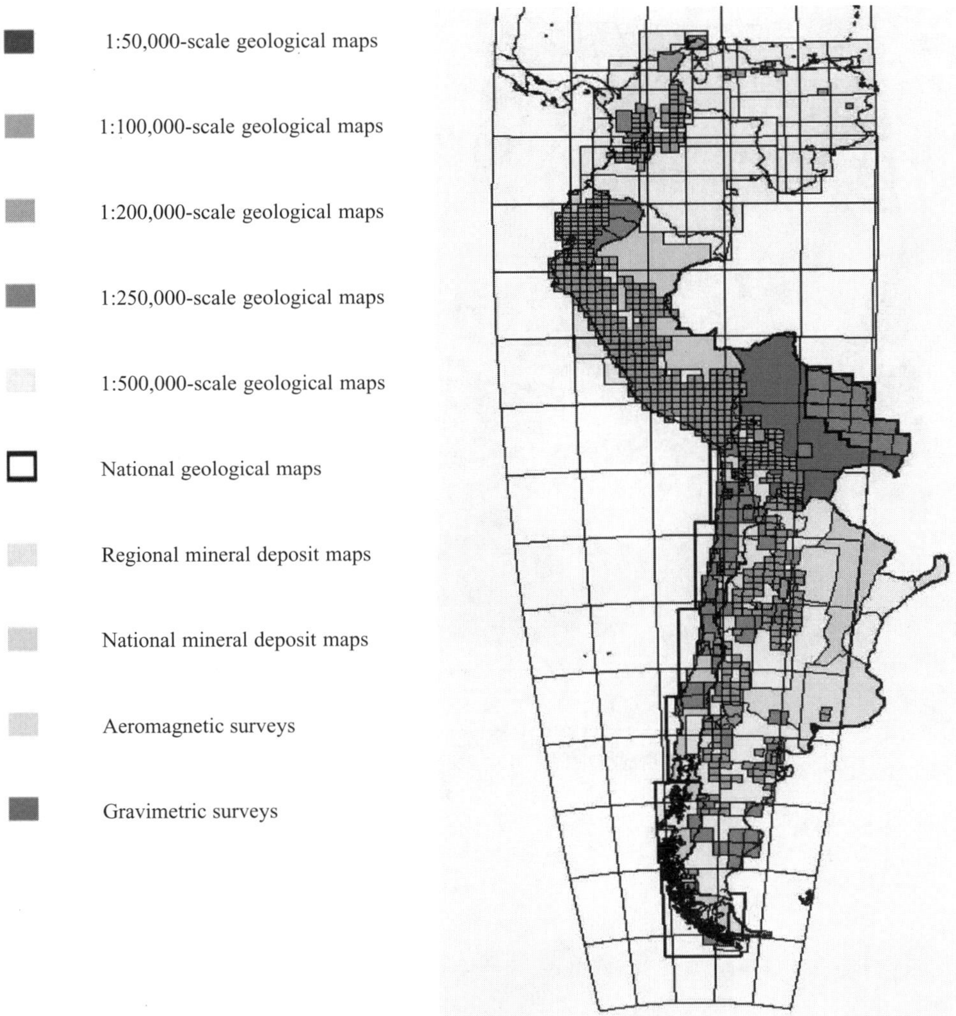

Figure 6: "Present state of the GIS Andes "Geological map coverage" layer.

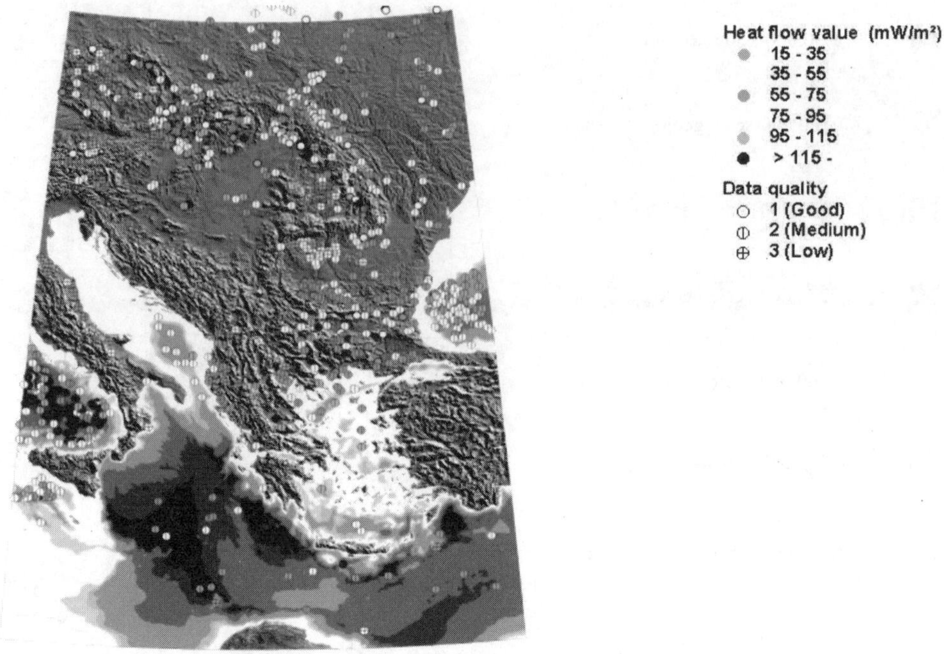

Figure 7a: GIS Central Europe "Heat flow" layer.

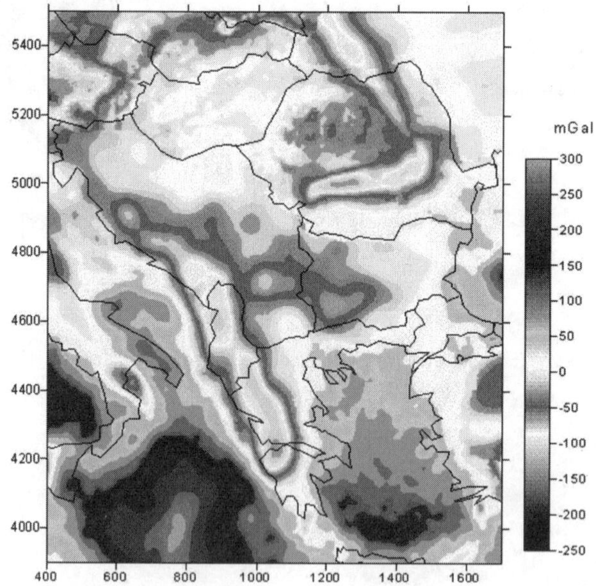

Figure 7b: GIS Central Europe "Gravimetric layer": composite Bouguer anomalies derived from gravimetric land data supplemented with satellite data.

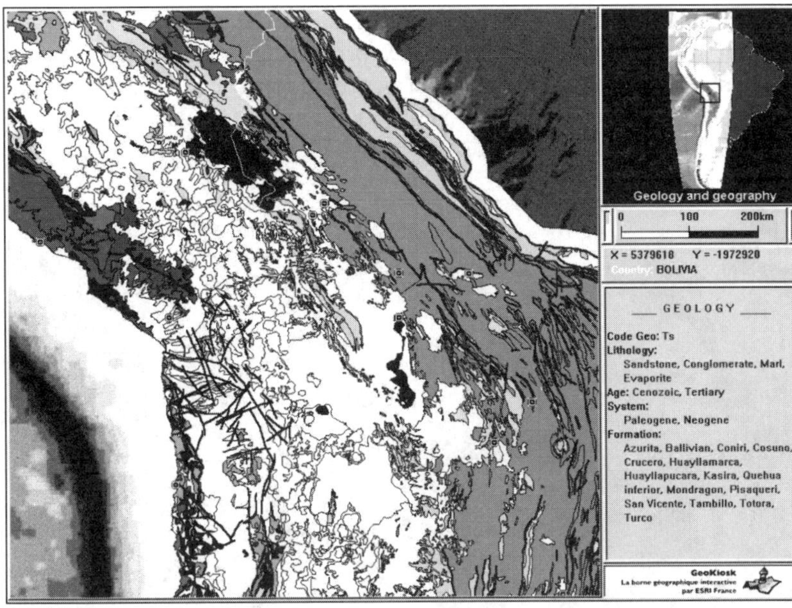

Figure 8a: Extract from GIS Andes (Cassard, 1999b [3]).

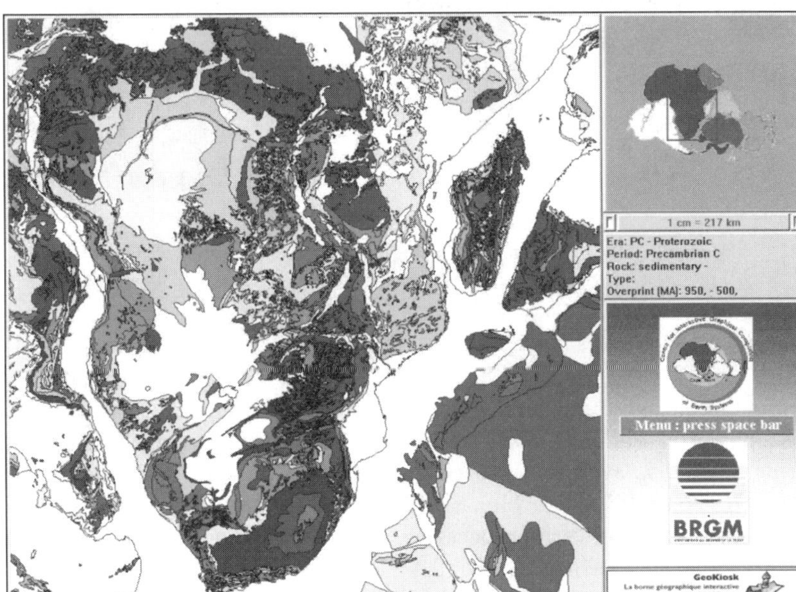

Figure 8b: Extract from GIS Gondwana (Nicol, 2000 [12]).

- Stage 4: Crossing the data between the various GIS layers so as to generate so-called "high added value" products. Such processing produces new data that in turn can be re-injected into the information system to form new layers — typical examples are Metallogenic potential or fertility maps, Predictive maps and Risk maps (Figure 9).

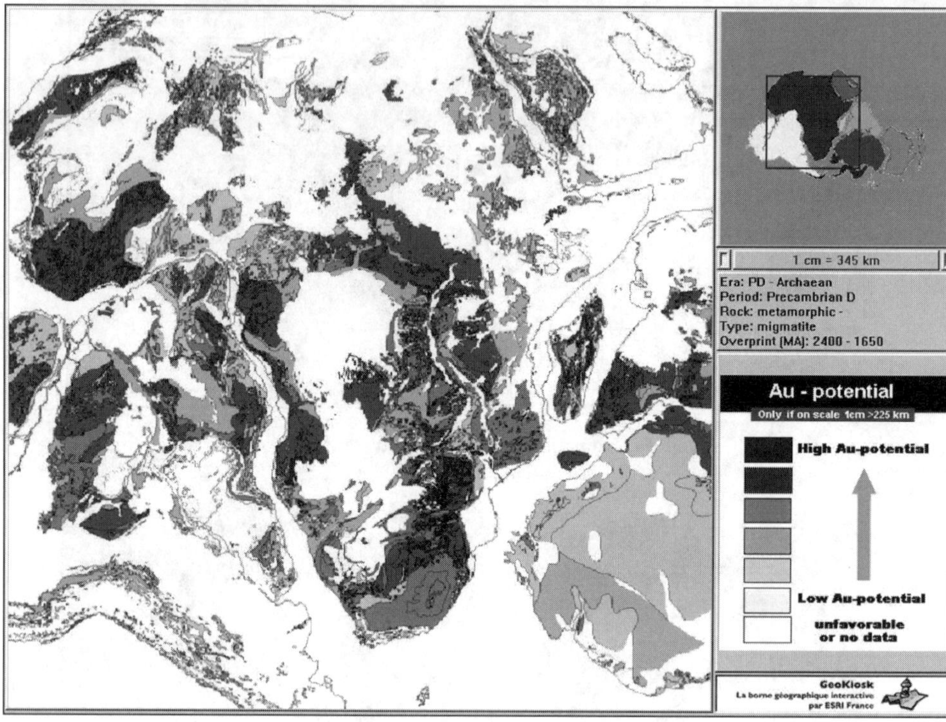

Figure 9: Gold potential map – Extract from GIS Gondwana (Nicol, 2000 [12]).

5. Using a GIS

Apart from its role as a digital archive (i.e. conservation of the geo-scientific portfolio), a GIS is a powerful work tool that gives ready access to the contained data and enables them to be easily displayed, superposed and crossed — it is the data crossing ability that makes the GIS so useful. To develop the full potential of a GIS, however, the system needs to be coupled to powerful processing tools that will enable a rational exploitation through several stages, such as:

- "Simple" superimposing. This allows one to graphically verify the spatial correlations between objects, but becomes increasingly difficult to implement as the number of data layers increases. Although this stage may be regarded as of limited interest insofar as it allows one to control associations that were already anticipated, it is nevertheless based on controlled and homogenized data and useful for determining scientific and/or inter-regional social problems using an approach that remains fundamentally naturalist;

- At a higher level, one can use model bases (e.g. the SignateX® base developed by BRGM for mineral deposits). These bases generally contain both standard and new models of association;
- The following stage could be to search the databases for new knowledge (KDD = Knowledge Discovery in Databases) using statistical analysis (e.g. Factor analysis of correspondences [FAC]) and Data Mining (extraction of information) (Figure 10);

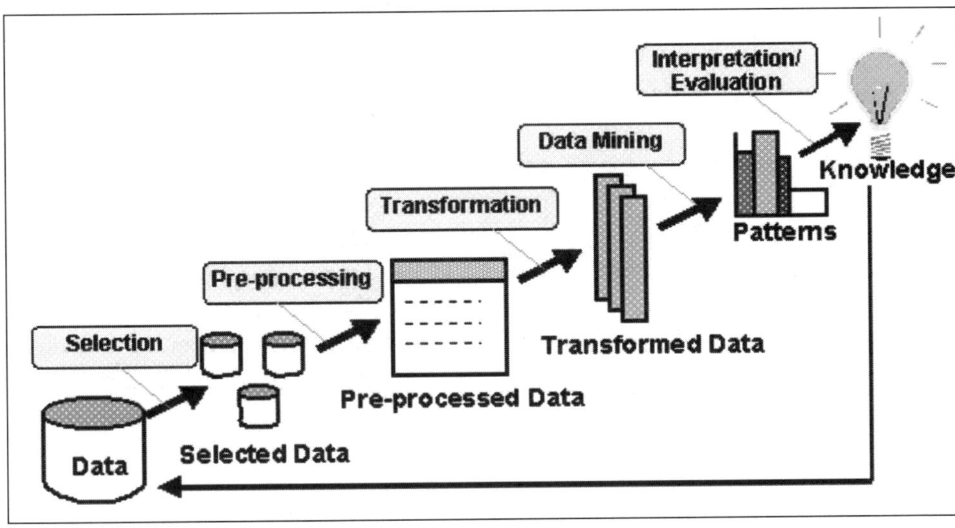

Figure 10: KDD process (after Fayyad et al., 1996 [7]).

Whereas classical statistical methods are essentially dedicated to numerical data, these more advanced techniques, and in particular Data Mining (which is relatively new in the domain of spatial databases), enable the integration of qualitative data and the determination of relational rules between objects. Data Mining (Salleb, 1999 [13]; Salleb and Vrain, 1999 [14]; Agrawal et al., 1993 [1]; Koperski and Han, 1995 [10]) makes it possible to qualify these rules (support and confidence) and also to treat text data (character strings); through its various tasks, it provides a better understanding and description of the data, enabling them to be classified (decision tree), associated (extracting association rules), compared and grouped (clustering). In addition, it can be supervised (i.e. targeted to a specific search) or not (blind extraction of the rules) whereupon it offers the possibility of discovering really new (not predictable) knowledge;

- The final stage may consist in combining the already formatted initial data (e.g.

interpolated or resulting from specific processing such as with "ERDAS®" for satellite imagery or "Geosoft®" for geophysical data, etc.) and/or new data (e.g. stemming from Data Mining). Various tools can be used during this stage: MapInfo®, ArcInfo®, ArcView®, etc. In general, the data that BRGM uses in its Metallogenic GIS are in ArcView format and the basic processing is done with ArcView coupled with Spatial Analyst®. For combining data, BRGM developed SynArc®, which is an extension of ArcView. SynArc allows multicriteria studies using the functions of similarity, classification, user-defined weighting from constructed multicriteria matrices, and probabilistic weighting (= weight of evidence modelling), and can also be used to test scenarios.

6. Prospects, conclusion

One of the aims in developing GIS Central Europe is to provide a powerful tool that may help answer current research questions in metallogeny, such as those listed by Heinrich and Neubauer (1999) [8], i.e.:

- Which processes at the scale of lithospheric plates control the development of a spatially and temporally restricted mineralized belt or mine district within a much larger complex collision zone such as the Alpine-Carpathian-Balkan chain? How does a regional explorer identify and assess the prospectivity of an entire mineral province?
- At the scale of a mineral province, what tectonic and thermal requirements are essential for creating a major ore-bearing structure or a fertile magmatic complex within a district of dominantly barren shears and plutons? Are there fundamental chemical and/or structural differences between fertile and barren magmas? How can the explorer identify such differences by geophysical methods or investigation of common barren samples? Is there a potential to identify structures controlling the locations of possible major ore deposits?
- At the scale of a hydrothermal system, what are the sources of ore metals, the extraction mechanism, the causes of fluid focussing and final ore-mineral precipitation? Which thermal, structural, and fluid- or magma-chemical processes determine the degree of enrichment of different metals, and thereby the localization, the size, the metal ratio and the bulk ore grade of an ore body? What are the mechanisms that lead to concentrated ore-mineral precipitation? How does the explorer find an economic ore body within a mineralized structure or intrusion?

With combined processing of the geographic, geological, mining and environmental data it will be possible to:

- Model hydric and atmospheric transfers;

- Characterize the environmental constraints to be incorporated in new mine development projects;
- Marry mine site rehabilitation projects to economic and environmental constraints;
- Resolve conflicts as to water and land distribution between the different economic activity sectors;
- Audit the monitoring network in terms of mining activity;
- Establish cross-border management of environmental aspects related to the industrial sector;
- Improve communication concerning these different themes by providing supporting maps.

GIS Central Europe will hopefully be a vehicle for geo-scientific communication and exchange between Geosurveys and Research Laboratories. It could, for example, serve to consolidate previous inter-regional syntheses and promote existing data (often of good quality, but underused because of being not updated and homogenized) to international decision-makers, investors, etc.
Finally, the ultimate aim of this GIS is to be able to produce decision-aid documents that integrate all the necessary factors related to industrial activities, and, in particular, the extractive industry, for medium- to long-term regional development planning and ecosystem protection.

7. References

1. Agrawal, R., Imielinski, T. and Swami, A. (1993) Mining Associations between Sets of Items in Massive Databases. *Proceedings of the ACM-SIGMOD 1993 International Conference on Management of Data, Washington D.C. (May 1993),* pp. 207-216.
2. Cassard, D. (1999a) Projet 00RESD02 – Synthèses Métallogéniques. BRGM Note REM/VADO, 16 p. (Unpublished).
3. Cassard, D (1999b) GIS Andes on the WEB: http://www.brgm.fr/sigand
4. Cassard, D. (1999c) GIS Andes: A metallogenic GIS of the Andes Cordillera. *4th International Symposium on Andean Geodynamics, (4.-6. October 1999), Göttingen*; *Extended Abstract,* Institut de Recherche pour le Développement Publ., (Paris 1999), pp. 147-150.
5. Cermak, V. (1993) Lithospheric thermal regimes in Europe. *Physics of the Earth and Planetary Interiors* **79**, 179-193.
6. de Boorder, H., Spakman, W., White, S.H. and Wortel, M.J.R. (1998) Late Cainozoic mineralization, orogenic collapse and slab detachment in the European Alpine Belt. *Earth and Planetary Science Letters* **164**, 569-575.
7. Fayyad, U.M., Smyth, G.P. and Uthurusamy, R. editors (1996) *Advances in Knowledge Discovery and Data Mining*, AAAI/MIT Press.

8. Heinrich, C. and Neubauer, F. (1999) ACB-GEODE, Geodynamics and Ore Deposit Evolution of the Alpine–Carpathian & Balkan Province – Status (August 1999). GEODE Internal Doc., 14 p. (Unpublished).
9. Jankovic, S. (1997) *The Carpatho-Balkanides and adjacent area: a sector of the Thethyan Eurasian metallogenic belt. Mineralium Deposita* **32**, 426-433.
10. Koperski, K. and Han, J. (1995) Discovery of Spatial Association Rules in Geographic Information Databases. *Proceedings of the 4th International Symposium on Large Spatial Databases (SSD95), Maine, (*Aug. 1995), pp. 47-66.
11. Nemcok, M., Pospisil, L., Lexa, J. and Donelick, R.A. (1998) *Tertiary subduction and slab break-off model of the Carpathian-Pannonian region. Tectonophysics* **295**, 307-340.
12. Nicol, N. (2000) GIS Gondwana on the WEB: http://gondwana.brgm.fr/
13. Salleb, A. (1999) Extraction des connaissances dans les BD géographiques. Rapport de stage de DEA. BRGM Report R 40803, 93 p. (Unpublished).
14. Salleb, A. and Vrain, C. (1999) *A tool for Mining Associations in Geographic Information Systems.* LIFO, University of Orléans, Paper ID 275, 20 p.
15. Tomek, C. and Pancardi Colleagues (1996) PANCARDI, Dynamics of Ongoing Orogeny, in Europrobe (1996) *Lithosphere Dynamics – Origin and Evolution of Continents.* European Science Foundation Publication, pp. 15-26.

MINERAL RESOURCE BASE OF THE CZECH REPUBLIC FOR THE XXI CENTURY AND ITS MANAGEMENT

P. KAVINA, J. STARÝ, M. VANĚČEK

KEYWORDS / ABSTRACT: mining production / import, export / mineral fuel sources / kaolin and clay deposits / feldspar deposits / silica sands deposits / graphite / limestones / building materials / legislative means of management / economic means / territorial planning.

Limited mineral fuel sources as coal deposits will be exhausted in a few decades. Czech Republic depends on import of crude oil, earth gas, metallic raw materials and raw materials for the production of fertilizers. Favourable conditions in raw materials for glass industry, ceramic industry and industrial use of kaolin fillers and pure limestone. All kinds of building materials are available. Territorial planning is considered to be the main tool of mineral resource base management.

* * *

Czech Republic belongs to the small countries, its area being 78,866 km^2, population about 10 millions, gross domestic product in 1999 – 53.1 billion USD. The share of industry in GDP was 38%, construction work share in GDP 4.9%, mineral raw materials share in GDP 1.8% (1999). 11,335 km^2 are covered by highly protected natural areas, which form 14.6% of the Czech territory, and in 19.3% of these areas mining is fully prohibited.

From the geological point of view the main structure is so-called Czech (Bohemian) Massif of the Variscan age with some sedimentary covert of Cretaceous, Tertiary and Quaternary ages. The easternmost part of the Czech Republic belongs to the Carpathian System of the Alpine age.

Coal deposits and various (relatively large) occurrences of non-metalliferous raw materials predominantly formed mineral raw material resources of the Czechoslovak Republic and later on, behind the split into two states in 1993 – into Czech and Slovak Republics – of the Czech Republic, too. Raw material base for metal industry was insufficient; no important crude oil and natural gas deposits were at disposal.

Decrease of the non-effective production, restructuring and modernization of industry with limitation of the share of the heavy industry in the gross domestic product (GDP) have been characteristic for the economic development in the Czech Republic after 1989. The production of the majority of mineral raw materials has decreased to 50 -

A.G. Tvalchrelidze and G. Morizot (eds.),
Mineral Resource Base of the Southern Caucasus and Systems for its Management in the XXI Century, 181–189.

80%. Import of iron ore, crude oil and raw materials for the production of fertilizers has decreased since 1989 as well.

Even when the domestic demand for mineral raw materials decreased, the damping of production of all the raw materials was not affected to the same extent. The domestic raw material producers took advantage of comparatively lower prices and a building boom in Germany for a higher export of building materials (cement, building stone) in 1991 and 1992. But due to the extreme energy-demanding industrial production, the decrease in the industrial production did not result in decrease of energy needs of the national economy. Irrespective of this fact, measures that could bring about the reduction of power consumption were not taken and the trend towards the building of other energy sources (atomic energy included) prevailed.

Exploitation of ores, fluorspar and barite deposits was terminated at the beginning of 90´s. Uranium ore mining has been gradually decreasing and coal production decreased too: 35 million tons of hard coal was mined in 1989 but there was only 17,2 million tons in 1999. The production of brown coal and high-volatile lignite decreased from nearly 90 million tons in 1989 to 45,4 million tons in 1999. The production of high percentage limestone also decreased from 8 million tons in 1989 to 4,7 million tons in 1999, that means the output decreased by 41.3%. Production of the crushed stone decreased by 47.9% and that of sand and gravel even by 65%. But for example feldspar output increased from 139 kt. to 244 kt. in the same period. Production of dimension stone increased from 543 kt. to 675 kt. and in 1998 even to 822 kt. There was mined more than 246 million tons of mineral raw materials in the Czech Republic in 1989, so the total output was 127 million tons in 1999 only. It means the decrease of the total mining production was nearly 49%.

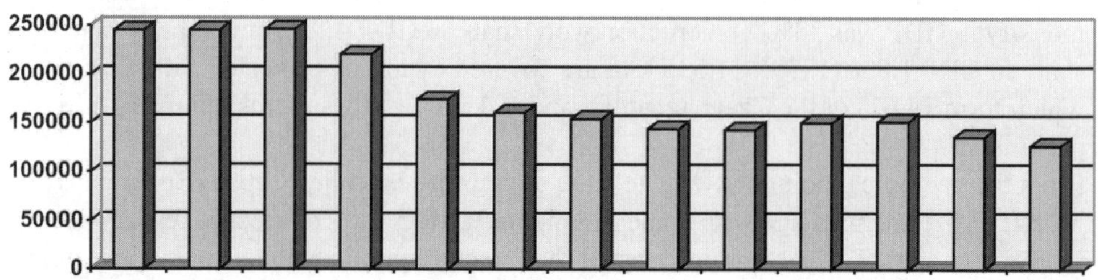

Figure 1. Total mining production – reserve deposits (kt.)

These changes in the use of domestic deposits of mineral raw materials naturally influenced the extent and the structure of import and export. The Czech Republic is completely dependent on the import of iron ores and metals, crude oil and natural gas, sulphur, salts and phosphoric raw materials.

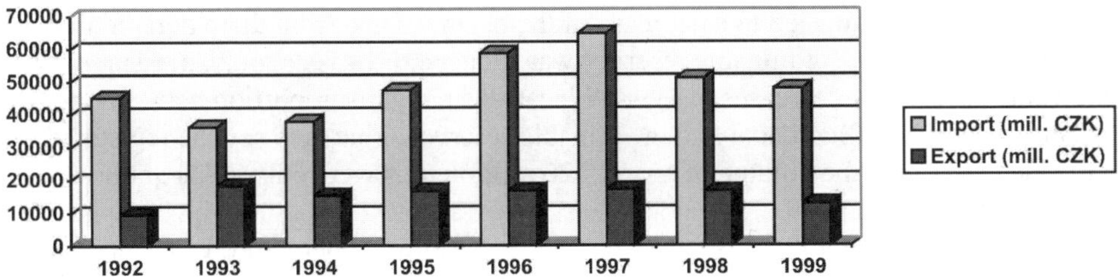

Figure 1. Raw materials in the Czech foreign trade

As concerns the import of mineral raw materials, it oscillated around 7 million tons of iron ores per year in the 90´s. About 7 million tons of crude oil per year were imported in 90´s, too. Natural gas import increased from 6,4 to 9,6 million cubic metres in the period between 1990-1998. The dependence on the import of fertilizers finds its expression in the import of 57.9 kt. of sulphur, 47.8 kt of phosphates, 31 kt. of phosphatic, 78 kt. of potassic and 68 kt. of phosphatic/potassic fertilizers in 1999.
The high degree of knowledge of the geological structure and exploration of the mineral raw material deposits are characteristic for the territory of the Czech Republic. Special emphasis was laid on the growth of the mineral raw material base. Considerable mineral raw material reserves were prospected and explored in the last period. The formal lifetime of the majority of raw materials has been still increasing in consequence of their production decrease.
On the other hand, changes of demand for the degree of demonstrated reserve knowledge occur along with the change of economic and technological conditions. That is why an extensive re-evaluation of the balanced reserves, that is to say, so-called „rebalance“ has been started under the management of GEOFOND. The rebalance will result in a real up-to-date evaluation of the reserves suitable for a potential industrial use and in an amortization of the reserves, the use of which will not come into question in the future.
Mineral raw material base of the Czech Republic can be divided into mineral fuel deposits, ore deposits, industrial mineral deposits and building material deposits. In this way the prospected and explored reserves of mineral raw materials are registered in the state balance. The mineral raw material base of the Czech Republic also includes the secondary raw materials of mineral origin, which are not treated in this paper.
Domestic primary energy sources are rather limited. It concerns especially extremely limited lifetime of the explored solid fuel deposits. Their enormously extensive production in the past, together with the high share of the primary energy sources brought our country to the present state. Even though the share of solid fuels in the energy balance is decreasing at the present time, there are problems with the complete use of present limited reserves, which are connected with the regional ecological limits. Both coking coal deposits and steam coal deposits are included in the balance of the hard coal reserves. Hard coal is mined practically only in the Upper Silesian basin. New conditions of use were created for this mining district in 1993. Increase of the minimum

thickness of coal seams led to a decrease of the balanced and from them derived minable reserves. The lifetime of minable reserves was reduced to 19 years only, measured by the value of reserves decreases in 1998. We can see a big disproportion between the balanced reserves of hard coal and the minable reserves. That is to say the registered minable reserves in the volume of nearly 720 million tons were only 7.7% of the balanced reserves in 1999.
Brown coal deposits still represent the main energy source. Energy producers envisage such a volume of brown coal exploitation that will ensure sources for production of about 50% of electric energy. This assumption, however, is in contradiction with the lifetime of minable brown coal reserves. There are about 1250 million tons of minable reserves at disposal in the mined deposits. Even if the production is decreasing because of the deposits' being mined out, the reserves will be used up till 2035. In addition more than 970 million tons are blocked by so-called area ecological limits imposed by the government, which could prolong life of the present reserves. The correction of the area limits, however, has not been taken into account in the government concept, as yet. The lifetime of brown coal reserves is as follows:

TABLE I. The lifetime of brown coal reserves

Reserves	Quantity (mill. tons)	Lifetime of reserves in years
Geological	9737	205
- mined	1881	40
- in exploited deposits	1351	28
- blocked by area limits	970	20

Note: lifetime is the share of reserves and their reduction by mining output

Natural gas and crude oil deposits have very limited extent and low reserves. It makes impossible to reach the production level that would substantially contribute to meeting the needs. Present-day production of domestic crude oil meets about 2.5% of the needs. Domestic production of natural gas meets the needs by about 1.4%. Even so, crude oil especially is an interesting complementary source used mainly in the production of frostproof greases. Besides that, abandoned deposits of liquid and gaseous hydrocarbons can be used as underground storage under favourable geological conditions, since their capacity on the territory of the Czech Republic is insufficient. Inasmuch as about 32% of state energy needs are dependent on crude oil and natural gas supplies, measures were taken for their diversification in the last years. So about 82% of crude oil was imported from Russia in 1999 and 18% was delivered by pipeline from Ingolstadt. Natural gas supplies have been diversified since 1997. Natural gas consumption of Norwegian origin will reach about 3 billion cubic metres per year after 2000. In 1999 78% of earth gas originated from Russia and 22% from Norway.
It is well known that U-ore deposits occur in the territory of the Czech Republic. Once

they represented the most important raw material base of the Comecon. The production of U-ores was subsidized by the state till 1993. The production of Uranium concentrate has decreased by a quarter since 1989, annual production of the last years being 600 tons of U. 2002 should terminate the exploitation of U-ores. Present consumption of uranium (in the Dukovany nuclear power station) reaches 330 tons per year. Annual consumption should increase up to 690 tons after starting both blocks of the new Temelín nuclear power station.

After the Second World War iron ores and nonf-errous metal mining developed. The output was subsidized in the form of the so-called limited costs that were often several times higher than the world prices. Level of these limited prices was decisive for demands of quality of prospected and explored iron ore deposits. With the exception of some Au-ore occurrences, the Czech Republic has no potential for effective use of ore deposits. That is why their mining activity was finished in 1993 and the explored reserves are systematically rebalanced.

On the other hand, a lot of industrial mineral and building material deposits represents effective production corresponding to present conditions. They supply raw materials for traditional Czech productions as glass, ceramic and paper industry. The building industry (production of mortar mixtures - lime, cement, calcined gypsum, building stone and bricks) is saturated with an adequate volume of raw materials, too.

Kaolin for china and porcelain production is mined in three deposits in the Karlovy Vary region and in a deposit in the Podbořany region and it has reached nearly 400 kt. annually in the last 5 or 6 years. Even though great reserves are registered and their life is about 70 years in mined deposits, two deposits in the Karlovy Vary region will be mined out in a short time. There is no equivalent substitute for it. About 50% of registered geological reserves are located in the Podbořany region. Their share in the production is just 30%. The lifetime of classical kaolin deposits from the Karlovy Vary region is only about 20 years. Production of kaolin used as fillers in paper and other industries (KP) is several times higher than production of kaolin for ceramics manufacturing. The main deposits of KP are situated in the Plzeň region, where 97% of the total KP production in the Czech Republic is mined in three deposits. Six deposits altogether are mined in the Czech Republic. The total production reached 4768 kt. in 1999. Even though kaolin exploitation is very extensive, the lifetime of balanced reserves is about 25 years in mined deposits. The total lifetime of balanced reserves is higher than 100 years in 22 KP deposits.

There are 120 various clay deposits registered in the Czech Republic, 24 of which are mined. Refractory clay deposits are situated mainly in these regions: Velké Opatovice, Nové Strašecí and Vyšehořovice. Clay for other refractory use (bond clay) is mined in the Cheb basin. Total production reached 636 kt. in 1999. About 30% of it were exported. For example, clays from the Cheb basin (whiteware clay, another refractory - bond clay, and clay for the production of floor tiles) have been exported to Germany for a long time.

Refractory clay for the production of chamotte is an important article of export. It is used for the production of refractory bricks, which are used for the walling of furnaces in metallurgy and glass industry. Clay reserves are in open deposits ensured for about 70 years. The lifetime of all deposits is estimated at several hundred years.

Feldspars are mined in five deposits and nepheline phonolites are mined only in one deposit. The production of feldspar was about 250 kt. in the last three years. The output of the production of phonolite was 24 kt. in 1999. Great volume of feldspars is mined in the form of feldspar-bearing sand in the Třeboň region. An important source is also albite aplitic granite from the deposit of Krásno. The lifetime of balanced reserves in mined deposits is about 60 years. High mining losses are not satisfactory. Sometimes they reached almost 70%. Another unfavourable factor is that more than 60% of the production is mined in deposits in protected natural areas. More than 20% of feldspar production is exported. It is interesting that the Czech Republic is not registered in the world statistics as a feldspar producer, although feldspar production reached 3.3% of the world production in 1999 and the Czech Republic would be the 8th among the main world producers.

Glass sand together with foundry sand is mined in three deposits in the Bohemian Cretaceous basin and in one deposit in the Tertiary Cheb basin. The production is about 1 million t./year Střeleč near Jičín is the most important deposit of glass sand in the Czech Republic. The lifetime of reserves in mined deposits is about 50 years, in other deposits 25-50 years.

The production of foundry sand was 717 kt. in 1999. Foundry sand is mined as an accompanying raw material (material of lower grade) in four mentioned deposits mainly of glass sand and they also form seven independent deposits in the Blansko and Svitavy regions. The lifetime of reserves in mined deposits is about 70 years and in registered deposits it is more than 180 years. The production of bentonite was 160 kt. in 1999. Bentonite is mined in three deposits in the territory of the Tertiary volcanites near Krušné hory Mountains. The lifetime of reserves is sufficient for more than one hundred years.

It is possible to say that graphite is the Czech national raw material. 15 deposits are registered in the balance of reserves, three of which were mined in 1999. Three of the registered deposits contain flaky graphite, one of them is mixed and others contain amorphous graphite. The total graphite production was 28 kt. in 1999. The lifetime of explored reserves in mining deposits is about 17 years only for flaky graphite and 33 years for amorphous graphite. The Czech Republic occupies the seventh place of the world statistics with the share of about 4.3% of the graphite world production, according to the Mineral Commodity Summaries of 1999.

In the Czech Republic there are favourable geological conditions for the occurrence of carbonate deposits. In our summary carbonate deposits are divided in pure limestone (limestone with very high percentage of $CaCO_3$), limestone used in cement production, and dolomite. 4673 kt. of pure limestone were mined in 1999. Since there is a lot of balanced reserves, pure limestone is a strategic commodity in context of the situation in the Central Europe. They can be used for hundreds of years, whereas in the neighbouring countries (except Slovakia) the lifetime of similar deposits is coming to an end. On the other hand, a considerable part of their reserves occurs in the protected natural area of Český Kras. Perspectives of pure limestone are mainly in use for the production of piece lime and piece limestone for chemical and food processing industries. They can be also used for ecological purposes. The production of other limestone, used for the production of cement was mainly 6.5 million tons in 1999. 13 limestone deposits for the production of cement are mined in the Czech Republic, their lifetime being more than 300 years. Clayey limestone for the production of cement is mined only in one deposit, the lifetime of which is about 50 years.

Production of stone used for building applications and production of architectural and sculpture dimension stone was annually about 300 000 cubic metres in the last few years. Blocks for architecture and sculpture are mined in 13 deposits; stone used in building applications is mined in other 70 deposits. Export of hard dimension stone and granite product balances import prices of soft dimension stone and slates. The lifetime of balanced reserves in mined deposits is hundreds of years but a part of them is situated in protected natural areas.

Even though building materials belong to the non-reserved minerals, prevailing part of them was industrially mined in the so-called reserved deposits. Production of building stone (crushed stone) has got settled on about 10 million cubic metres in the last years. 9442 thousand cubic metres were mined in reserve deposits and about 850 thousand cubic metres were mined in non-reserve deposits in 1999. About 180 reserve deposits and about 50 non-reserve deposits were exploited. The deposits are located practically on the whole territory of the Bohemian Massif (except the territory of the Bohemian Cretaceous basin). Just a few of them are situated in the Western Carpathians. Some of the deposits are located in the protected natural area of České středohoří. Mining production of sand and gravel reached 8181 thousand cubic metres in reserve deposits and it was more than 4600 thousand cubic metres in non-reserve deposits in 1999. This volume was mined in about 80 reserve deposits and in more than one hundred non-reserve ones. Many important deposits are located in catchment areas of big rivers. High volume of the exploitation is carried out in the protected natural area of Třeboňsko.

The range of brick elements and roof tiles in brickmaking industry has been extended thanks to foreign investment after 1989. Production capacities and quality of products have increased at the same time. Production of perforated bricks with high heat resistance was newly opened. The production of brick clays has been oscillating around 2 million cubic metres per year since 1992. 1934 thousand cubic metres were mined in reserve deposits and more than 60 thousand cubic metres in non-reserve deposits in 1999. Even though more than 60 mined deposits occur in the territory of the Czech Republic, there are problems with their uneven regional distribution.

At the present time state enjoys just limited means to influence the management of mineral raw materials. These are: means of legislative character, economical means and territorial planning.

The legislative means arise from the Mining Law, the Act on Mining Operations (Activity), the Act on Geological Works, the Natural Protection Law and the Building Law.

The Czech Mining Law distinguishes between reserve and non-reserve minerals.

Reserve mineral raw materials are the following: mineral fuels, metallic ores and a part of industrial minerals. Deposits of many types of dimension stone are reserve, too. Other building materials are non-reserve minerals. If they were mined to an industrial extent in the past, they were declared reserve deposits and were handled accordingly. Right of prospection, exploration and mining of reserve minerals is granted by the state institutions in an administrative procedure. Territorial decision and authorization of the activity by mining method are required for the beginning of mining the deposits of non-reserve minerals, which are integral part of the land. Relevant Building and Regional Mining Offices issue these documents. The Czech Mining Law enables economical use of mineral resources, requires their protection, the total extraction of reserves and minimum losses. The Act on Geological Works is being amended presently. Geological works are divided into non-destructive works and works which destroy lands. Functional use of mineral resources follows the Law on Natural Protection and the Building Law. The Building Law refers also to the principles of territorial planning. Options of territorial planning documentation are the following:

- to exercise the protected areas of known mineral deposits.
- to define areas, where mining activity cannot be carried out or it is limited by conflicts of interests.
- to define chronology of exploitation of mineral deposits in the area.
- to define conditions for clean-up and revitalization of the area after finishing mining.

Economic means in function are the following:

- Annual fee for the mining claim area of 10 000 CZK for each km^2. Regional Mining Office transfers this fee into the budget of the municipality within whose territory the mining claim is located.
- Annual fee for the extracted reserve minerals up to 10% of its marked price. The Regional Mining Office transfers 50% of the income from fees into the state budget and 50% into the budget of the municipality within whose territory the mining claim is located.
- Financial reserve from costs used for settlement of mining damages and revitalization.
- 10 % amortization from price of material property in case of purchase of expensive investments for mining activity and processing.

The entrepreneur's obligation to pay back all the funds for prospecting and exploration that came from the state budget in the past is not stimulating.

Following new economical means are being considered:

- Ecological fee.
- Resources depletion allowance.
- Financial advantages for technologies with low or without wastes.

An adequate informational system is necessary in order to create a sensible raw material politics of the state. GEOFOND is in charge of gathering and processing the information on mineral raw materials in the Czech Republic. This information system documents the situation about mineral raw materials not only in the Czech Republic, but also in foreign countries. Another task is to lead agenda of geological and mining works permits. GEOFOND runs an official balance of prospected and explored mineral raw material reserves, publishes “Mineral Commodity Summaries of the Czech Republic” also in English and issues “Summaries of Reserves Change in Registered Mineral Raw Material Deposits” every year.

SUBJECT INDEX

6. Albania – geological premises / mineral resources potential of Albania / mineral production / legislation of mining sector / mining taxation / institutions in mining sector / Albanian know-how in mining industry / privatization of mining industry of Albania / privatization targets / private investments in mining industry / opportunities and incentives for foreign investors / specific categories of risk covered by Political Risk Guarantee Facility / market considerations / construction materials market / chrome market. G. Demi et al.

7. Ancient tradition of mining in Greece / more recent history of mining in Greece / main exploitation centres / the main industrial minerals / Greece as a country, possessing the only commercially exploited deposit of huntite in the world / Greek marbles / main applications of Greek mineral products / key points in XXI century mining / sustainable growth and industrial minerals / proposals. M. Grossou-Valta, F.Chalkiopoulou.

8. Classification of Portuguese mineral resources in subsectors / metallic ores / decrease in their importance both in Portugal and in the world / three major deposits of metallic ores – Neves-Corvo, Aljustrel and Panasqueira / subsector of non-metallic ores / industrial minerals for the ceramic industry / non-traditional new resources – arkoses and spodumene / industrial stones: sands and building materials / ornamental stones: granites, marbles, and limestones / rising tendency in these two subsectors / great potential of natural mineral and spring waters / the tendency to increase shown by the subsector in the last few years. M. L. P. de Amorim Dinis, A. Botelho de Sousa.

9. Central and south-eastern Europe / GIS / metallogeny / ore deposits / metal potential and fertility maps / geodynamics / environment / risk maps / D. Cassard, Y. Itard.

10. Mining production / import, export / mineral fuel sources / kaolin and clay deposits / feldspar deposits / silica sands deposits / graphite / limestones / building materials / legislative means of management / economic means / territorial planning / M. Vaněček et al.